The Power of BOLDNESS

Ten Master Builders of American Industry Tell Their Success Stories

Edited by Elkan Blout

JOSEPH HENRY PRESS
Washington, D.C. 1996

JOSEPH HENRY PRESS • 2101 Constitution Ave., N.W. • Washington, DC 20418

The Joseph Henry Press, an imprint of the National Academy Press, was created with the goal of making books on science, technology, and health more widely available to professionals and the public. Joseph Henry was one of the founders of the National Academy of Sciences and a leader of early American science.

Library of Congress Cataloging-in-Publication Data

The power of boldness : ten master builders of American industry tell
  their success stories / edited by Elkan Blout.
      p.   cm.
    Includes index.
    ISBN 0-309-05446-X (hardcover : alk. paper). — ISBN 0-309-05445-1
(paperback : alk. paper)
    1. Businessmen—United States—Biography. 2. Industrialists—
United States—Biography. 3. Inventors—United States—Biography.
4. Entrepreneurship—United States—Case studies. 5. Family-owned
business enterprises—United States—Case studies. 6. Industries—
United States—History.   I. Blout, Elkan.
HC102.5.A2P69   1996
338.092'273—dc20
  [B]   96-31131
                                                             CIP

The Power of Boldness first appeared in the Spring 1996 issue of *Daedalus,* the journal of the American Academy of Arts and Sciences.

Printed in the United States of America

Preface

ONE COLD WINTER MORNING four years ago when I was talking to myself, the idea for this book was born. The genesis was really my personal ruminations about the state of American society—both positive and negative. I felt that one of the main contributions our society had made to the world of the twentieth century was its emphasis on and rewards for initiatives and inventions. I believed that this subject was not fully understood and appreciated by many elements of our society. In addition, in 1992 it was not clear that American industry would be able to resume its leadership in a very competitive world economy.

Out of these thoughts arose the idea of a book written by those people in American industry who had made major contributions to the shaping of the industry during the last several decades. At this point I thought about the American Academy of Arts and Sciences and its distinguished and diverse membership, in both the academic and industrial worlds, and wondered whether the Academy might be an appropriate venue and sponsor for such an activity. I brought the embryonic suggestion of this book, which I called at that time *American Industry at the Crossroads,* to the attention of the Council at its next meeting. The Academy's Council gave me its enthusiastic support for pursuing the idea, and then I seriously began to work on it.

My background was in industry, in academic science at Harvard University, and for the last twenty years as an officer and council member of the National Academy of Sciences and the American Academy of Arts and Sciences. With this background I had a fairly large acquaintance with the scientific, technological, and industrial leaders in the United States. I was heartened to proceed when I broached the idea to a couple of them and received enthusiastic and positive responses.

The next step was to approach a sponsor for the financial support of the book. An application to the Alfred P. Sloan Foundation resulted in an important, initial grant in May 1993, and this was soon followed by grants from two other private foundations (the Cabot Family Charitable Trust and a source that wishes to remain anonymous). This support, for which I am grateful, allowed me to firm up my ideas, to recruit authors and, most importantly, an editor.

I am very pleased that almost all the people I asked to write essays responded affirmatively and often enthusiastically. In my discussions with potential authors I emphasized that we were not looking for theories from scholars and economists, but rather we wished to emphasize stories from people who built America into what it is today. I also emphasized that any essay should be a personal account and, I hoped, would indicate the author's opinion about influential people in his career and other circumstances that molded him into the leadership position he held in the technological industry—the backbone of America's social and economic position in the world.

As you read the essays in this volume you will note that the authors have, indeed, heeded my suggestions. They are all very different and, for the most part, very personal. Not all the people who agreed to contribute to this volume did eventually write an essay. I learned that there are many reasons for people not to follow through after initially agreeing to participate in such an endeavor. I hope the appearance of this volume may stimulate a few of our potential "laggard" authors so that we may have the benefit of their insights in the future.

Finally, I would like to express my gratitude to my tireless and thoughtful editor, Upton Brady. Upton Brady was formerly the Director of the Atlantic Monthly Press and brought to his position in this endeavor a wealth of knowledge about writing, about authors, and about publishing. All this knowledge proved most useful over the past three years, and I very much appreciate his sensitivity and willingness to undertake difficult tasks as this book grew. Also, I want to specify my loving appreciation and gratitude to my wife, Gail, who suffered with me through some of the crises I experienced during this project. As usual, she was a solid rock of common sense, thoughtfulness, and support.

I am very grateful that my friend and colleague, Professor Alfred Chandler, agreed to write the introductory essay that puts the subsequent essays into perspective in a masterful manner. These essays appeared earlier this year in *Dædalus*, the Journal of the American Academy of Arts and Sciences, whose editor, Stephen R. Graubard, and associate editor, Phyllis S. Bendell, contributed a lot to the final product. I learned much during these last four years, and I hope that this volume may indeed have an influence on younger Americans as well as those in our society who are devoted to politics and government.

Elkan Blout
Boston, Massachusetts

Contents

The Power of BOLDNESS

Alfred D. Chandler, Jr.

Introduction: Entrepreneurial Achievements

T HIS VOLUME IS COMPRISED of a disparate set of essays written by ten business leaders of different backgrounds and different ages. Their birth dates range from 1897 to 1962. The essays vary in their approach, content, and tone. Some deal with the histories of their enterprise; others focus more on the underlying elements of business performance and values. But all have a common core. All ten men built or enlarged highly successful business companies—enterprises that brought them substantial personal wealth, increased the competitive strength of their industries, and added to the productivity and growth of the nation's economy.

To help the reader appraise and evaluate these essays, this introductory chapter briefly reviews the education, training, and impressive entrepreneurial achievements of each of the men involved. Their careers were of two types. Six were technological inventors who created the enterprises needed to commercialize their inventions. The other four took over the management of their fathers' enterprises and led their companies into new industrial and geographical markets. This review not only assists in the evaluation of the individual entrepreneurs but also provides insight into the processes of entrepreneurial achievement in the United States during the years since World War II and the impact of such entrepreneurial activities on the nation's industrial competitive strength.

Alfred D. Chandler, Jr. is Straus Professor of Business History, Emeritus, at Harvard Business School.

THE INVENTOR-ENTREPRENEURS

John F. Taplin, born in Massachusetts in 1913, represents the prototype of the inventor-entrepreneur. From his earliest youth he was intrigued by things mechanical and saw himself as an inventor. A member of the class of 1935 at MIT, he received a solid foundation in electrical engineering and displayed special competence in automatic feedback systems. He then spent three years getting hands-on experience as a toolmaker for the Foxboro Company, a maker of complex instruments. After the outbreak of World War II, he went to another company as an engine designer. There he invented a self-contained vacuum governor that was used to control the speed of engines driving electric generators in aircraft. In 1943 he joined the Fenwal Company to produce plastic bags to transport human blood and one of the first dialysis machines that was used as a temporary artificial kidney.

In 1946, on the basis of this experience, Taplin started his own enterprise, the Kendall Controls Company, to develop "superaccurate" pressure regulators. These regulators were sold to his first employer, the Foxboro Company, who resold them through their worldwide sales force. A lucrative licensing agreement soon followed with the Sheffield Air Gauge Company, the producer of over 50 percent of the air gauges made in the United States. With these orders in hand, Taplin used retained earnings to build a large production plant and a national sales organization with sales of $25 million. Next came the large-scale production of one of the internal parts of the regulators, a rubber and fabric rolling diaphragm seal, in a new plant built on Route 128 in Massachusetts, a change of the company's name to Bellofram Corporation, and an increase in the number of sales offices in the United States as well as licenses sold to foreign companies. Again, growth was funded by retained earnings.

In 1965 Taplin realized that he preferred to be an inventor rather than a CEO. He turned the management of his enterprise over to "an experienced professional president and chief executive officer" so that he could devote his time to designing new products. In addition to his work at Bellofram, he developed an automatic reading machine. After it had been perfected and had become a several million dollar business, he sold it to the Burroughs

Company, one of the nation's largest business machine companies, to automatically read and sort checks cashed by bank customers. Shortly thereafter Taplin decided to sell the Bellofram Corporation, having concluded that "the untapped future sales potential of the unique, patented products in Bellofram's inventory" could best be realized by "merging with a firm that had a global sales organization producing and marketing somewhat similar industrial products."

Whereas Taplin decided to remain an inventor, Ralph Landau continued on as a CEO. Landau and Taplin overlapped as undergraduates at MIT: Landau in chemical engineering, Taplin in electrical engineering. Landau stayed on and received his Doctor of Science degree in 1941. During the war he "learned the art of chemical engineering process design" from his work at the M. W. Kellogg Company, a large research, engineering, and construction contractor, primarily for oil companies. This work placed him at the forefront of the revolutionary developments in the new products and processes of polymer petrochemical technologies that so transformed the chemical industry during and after World War II.

In 1946 Landau cofounded Scientific Design Company, Inc. with Harry A. Rehnberg (it later became Halcon International and Halcon SD Group). Theirs was a research company that "performed contract design and engineering services for our clients' technology," and they invested their resulting fees into their own research. The recruitment of a "unique team of brilliant technologists" resulted in the invention and licensing of more than a dozen innovations in petrochemical products and processes between 1946 and 1966. From the start, however, the partners aspired to be a chemical manufacturer using their own technology. That was where the profits lay, particularly as they had learned that "licensees were increasingly willing to pay only for our successes and not our failures."

Such profitable manufacturing operated on a far greater scale and called for a larger amount of financing than did Taplin's specialized instruments businesses. The response to this challenge was a partnership with a large oil company, ARCO (Atlantic Richfield Company), which at that time was moving into petrochemicals. In 1966, Halcon and ARCO formed a 50/50 joint venture, the Oxirane Corporation, to produce propylene oxide, the principal ingredient in urethane foams and other polyurethane polymers. Halcon provided the technology and ARCO contributed 90 percent of the initial funding, with a repayment schedule that took 50 percent of the joint ventures' cash flow.

Because the Halcon process had far lower costs than the older chlorohydrin process, Oxirane quickly gained substantial market share. The first plant came on stream in Bayport, Texas in January 1969, followed by a second plant and then a third one at Bayport, one in Rotterdam, a joint venture to build facilities in Japan, and a similar venture to build in Spain. The process of designing, scaling up, and building the plants was in itself a classic example of technological improvement through continual "learning by doing." By 1980 Oxirane's revenues had reached the billion dollar level, and Landau, the inventor-entrepreneur, was becoming an experienced CEO.

Nevertheless, this technological achievement was soon threatened by external conditions. The company had financed a huge new plant in Channelview, Texas. Commissioned in 1978, shortly before the second oil crisis, it was financed outside of the original agreement with ARCO, primarily because ARCO was then investing heavily in the recently opened Alaskan fields. Inflation was rising rapidly. As the price of raw materials soared, so did interest rates, reaching 21 percent. "Suddenly, all our cash flow was going to the banks," prompting Landau to sell Halcon's interest in Oxirane to ARCO. As oil prices declined, inflation receded, and ARCO expanded the operations at Channelview and Rotterdam, built a new plant in France, and became the world's leading producer in this market.

By 1992 Halcon had sold off its remaining engineering licensing and other businesses. Landau became a member of the economics community at Stanford and Harvard, where he studied the interaction between technology and economics; he presents valuable observations on this in the concluding part of his essay.

Edwin ("Din") Land, slightly older than Landau, was, like Taplin, a born inventor. Research as a Harvard undergraduate led him to leave Harvard before graduation so that he could concentrate on the development of a plastic sheet for polarizing light. Once he was successful, he formed Land-Wheelwright Laboratories in 1933 to continue research on applications for his invention. Then, in 1937, with financing from Wall Street, he established the Polaroid Corporation to commercialize a plastic polarizer to reduce headlight glare, work that continued throughout World War II. During the war, however, Polaroid concentrated on government contracts to develop new products, including specialized optical filters, plastic optics, vectographs for aerial photo interpretation, infrared night adaptation goggles, and even a heat-seeking "smart" aerial bomb to sink naval vessels. At the war's end Polaroid employed more than nine hundred workers and had revenues of $16 million; by 1946 employment had dropped to about 250 and the company was $800,000 in the red. Most of the government contract work had been sold off to Eastman Kodak. Moreover, the automobile industry had failed to adopt Polaroid's headlight system.

But even before the war's end, Land had begun work on his "dream" of a small, hand-held, easily operated camera that would produce and finish dry photographs in full color. By 1946 Land had defined the basic photographic process. During the next two years the small research company concentrated on developing a commercially viable product. In 1948 the world's first "instant" camera came on the market. That year the company lost $866,000 on $1.4 million in sales. In 1949 it showed a profit before taxes of $770,000 on $7.0 million in sales.

In the 1950s the company made the initial large-scale investment in manufacturing, marketing, and research and development necessary to reach and maintain national and global markets. In these moves a senior executive, William McCune, led the way in "providing the vision and infrastructure in product development and manufacturing." In 1954 a large plant in Waltham, Massachusetts was built to produce black-and-white film. Between 1952 and 1958 the research group, working closely with researchers at Eastman Kodak, concentrated on the development of color film. By 1957, when Polaroid's stock was listed on the New York Stock Exchange, sales were over $50 million. By 1960 its net sales had reached almost $100 million, and Din Land had become the CEO of a major US industrial corporation; he also remained the company's foremost inventor.

In 1967 Land turned to completing the final step in achieving his dream, namely, an integrated automatic camera. He took full charge of the research for the new product, the SX-70. It was a massive technological effort, costing over half a billion dollars (the financing came from retained earnings). It was not finally introduced until 1972. Some of the new technologies worked well, others did not. Sales were disappointingly slow, but they were still substantial.

During these years Land continued to dominate the company by remaining Chairman, CEO, and Director of Research. As he put it: "There isn't any number two; there are a lot of number threes," whom he sometimes played against each other. He continued to concentrate on product development, paying much less attention to day-to-day management, finance, managerial recruitment, or the development of closely related product lines.

By the late-1970s this style of management was affecting the company's performance. In 1976, Eastman Kodak came out with a high quality, less expensive version of the SX-70. (Eastman later paid Polaroid over $900 million for patent infringement.) At the same time, Land was moving forward to develop an instant movie camera, which was introduced in 1977. Again an impressive technical achievement, but its sales were disappointing. With intensified competition and less than impressive product development, the years between 1978 and 1983 were the most difficult in Polaroid's history. In 1980 Land turned the position of CEO over to McCune, and in 1982 he retired as Chairman of the Board and Director of Research.

Land's career as an inventor-entrepreneur was brilliant; his career as a manager was less so. This phenomenon has been common in American industrial and business history from George Westinghouse to Henry Ford to Steven Jobs.

An impressive exception to the proposition that inventor-entrepreneurs tend to be less than effective managers is the career of Gordon E. Moore. The reason may be that from the start he was one of a team of three: himself, Robert N. Noyce, and Andrew S. Grove. This team was at the forefront of commercializing what became probably the most revolutionary technological innovation of the twentieth century, the semiconductor—first the electronic transistor and then the integrated chip. (One member of the team, Noyce, patented the integrated chip at Fairchild Semiconductor in 1959 at almost the same moment that Jack Kilby at Texas Instruments did.) Fairchild Semiconductor was born in 1957 when

Moore, Noyce, and six other scientists and engineers left Shockley Semiconductor Laboratory, which was formed in 1954 near Palo Alto by one of the inventors of the transistor. At Fairchild, Noyce, a graduate of MIT with a Ph.D. in 1953, became vice president and general manager. Moore, who received his Ph.D. in chemistry and physics in 1952 from the California Institute of Technology, became its director of research and development, and Andrew Grove, who joined Fairchild in 1963, the year he received his Ph.D. from the University of California, Berkeley, became Moore's assistant director. Moore's challenge at Fairchild was not so much inventing as it was getting the innovations manufactured and on the market before spin-offs from Fairchild did so. The problem was that his laboratory was "spawning new companies rather than contributing to the growth of Fairchild."

So in 1969 the three left Fairchild to form the Intel Corporation. Financed in large part by a venture capitalist, Arthur Rock, its purpose was to manufacture in volume complex memory chips based on the Metal Oxide Semiconductor (MOS) technology developed at Fairchild. Moore and Grove concentrated on development, Noyce, as CEO, on management. Their first product was a 64-bit static Random Access Memory (RAM). The commercial success of this high speed memory identified Intel as more than a typical Silicon Valley research firm. Next came a more versatile and powerful memory chip, the D (for dynamic) RAM chip, followed by the more advanced Erasable Programmable Read Only Memory (EPROM). To expand their market the team created a Memory Systems Division that sold complete memory systems worldwide, forcing the firm to increase its engineering and marketing capabilities. By 1974 Intel had firmly established itself with a 40 percent pretax profit.

Then, in the late 1970s, large Japanese enterprises with the technology to manufacture high quality DRAM chips in great volume with lower unit costs and prices began to move into the American market in strength. By the early 1980s Intel and other American producers, unable to compete, were being forced to shut down their plants.

Fortunately these chip producers were saved by the invention of the microprocessor. "A computer on a chip," its computing power was far greater than existing computing (as opposed to memory) chips. Indeed, Intel had been a pioneer in the development of the microprocessor in the early 1970s. But for most of the decade it was used primarily in dedicated control systems (as a microcontroller) in automobiles, appliances, automated production lines, and the like. During the second half of the 1970s Intel's revenues from its Memory Systems Division were much greater than those from its microprocessors.

It was IBM's brilliant move into personal computers that transformed the computer industry and, as Moore points out, "changed the entire course of Intel's history." In 1980 IBM set up a unit in Boca Raton, Florida, to design a personal computer and build a mass production plant and a global marketing organization, all within a year. To meet the deadline, the unit managers had to obtain their components and software from outside suppliers. They chose the Intel 8088 as their chip and William Gates's DOS as their software. In the third full year of operation, revenues of that IBM unit were close to $4 billion. IBM clones poured into the market. After a shakeout in 1985 and 1986, brought on by the resulting excess capacity, the market recovered.

The clones, such as Compaq Computer, rapidly took market share from IBM. But nearly all had to use Intel's chips and Microsoft's software. In addition to giving Intel (and Gates) probably the most valuable franchise in American industrial history, IBM provided it with $400 million, which tided Intel over the financial stringency resulting from the loss of the memory chip business and the cost of designing and constructing the new microprocessor plants. It did so by purchasing 20 percent of Intel's stock, which it later sold off.

With the huge and growing market assured by the IBM franchise, Intel quickly became the world's largest producer of microprocessors. By the mid-1990s it was enjoying $10 billion in revenues. Although IBM gave Intel the opportunity, Intel's success in exploiting it reflects the extraordinary, technologically sophisticated organizational capabilities that Moore, Noyce, and Grove had built.

George N. Hatsopoulos and William M. Haney, III are members of a younger generation of technological entrepreneurs. Hatsopoulos spent World War II in his homeland, Greece. After coming to the United States he entered MIT, receiving his advanced degree in 1956. Haney graduated from Harvard in 1984. Their career patterns were similar to those of the inventor-entrepreneurs just described. They initially built research organizations and then formed their own enterprises to manufacture and market their inventions. But they operated in a different environment—one that had been shaped by the products of the chemical and electronic revolutions in which Landau and Moore had played a part, and one that reflected both environmental and regulatory changes of the postwar world.

George Hatsopoulos's career began with his doctoral dissertation at MIT. Using thermodynamic principles, he devised a theoretical engine that converted heat directly into electricity without using any moving parts. With a young partner then at the Harvard Business School, he formed Thermo Electron to build a prototype and achieve proof of his theory. Unable to find a commercial application for the engine, he used the acquired knowledge to do specialized research work, with contracts coming largely from government agencies and gas utilities. In 1966, when revenues had reached about $2 million, the partnership "went public to raise capital for continued R&D and to set up operations for commercializing our first products" by obtaining the necessary production facilities and marketing channels.

The Clean Air Act of 1970 provided the opportunity to develop a new product line—instruments to measure pollutant emissions. A large order from the Ford Motor Company established the company as the first major producer of such an instrument. As a result, the EPA designated Thermo Electron's technology as the nation's standard, and orders poured in from automobile manufacturers from around the world.

The energy crisis that began with the OPEC oil embargo in 1973 provided an even broader opportunity for developing new energy-efficient equipment, particularly for the paper and metallurgy industries. Because of the company's technological expertise, it was asked to contribute to the Public Utilities Regulatory Policy Act of 1978. That act and the soaring oil prices after the Iranian Revolution led to the commercialization of such products as industrial heat-recovery systems and cogeneration equipment for the production of both electricity and process heat from the same fuel source.

Then, with the sharp recession of the early 1980s and the rapid decline in oil prices, Thermo Electron found itself "without a market—our customer base had simply disappeared." At the same time, the earnings that had funded the company's research also disappeared. Hatsopoulos was faced with the choice of either maintaining research or maintaining dividends. He took the former, accepting the risk that would come with the resulting drop in the value of Thermo Electron's stock at a time when predatory corporate raiders were searching for strong firms at bargain prices.

Hatsopoulos, who had written a book, *High Cost of Capital: Handicap of American Industry,* responded to this challenge with an innovative financial strategy. He would "spin out" the company's development projects as publicly owned subsidiaries in which the parent held the large majority of the stock. Besides providing needed funding for continuing R&D, this move had the additional advantage of permitting the development team to receive equity in the spinout and thus have a stake of their own in the project's success. The first such spinout came in June 1983 with the formation of Thermedics, with the parent company retaining 84 percent of its equity. Other spinouts followed. By the early 1990s Thermo Electron, by then a billion dollar company, consisted of a dozen company-controlled subsidiaries that included ThermoTrex, the company's R&D organization. The parent remained responsible for strategic direction, financial resources, and administrative services ranging from legal to human resources. The spin-offs implemented strategies agreed upon by the parent. "Energy and instrumentation are the primary foundations upon which virtually all these businesses have been built," and the unifying bond remains technological expertise. In this way, George Hatsopoulos has combined technical knowledge and corporate managerial skills to create a new type of a billion dollar company well suited to utilize US technological capabilities into the next century.

William Haney's career is only beginning, but his initial steps follow the familiar pattern. As Din Land did forty years before, he became an inventor while he was still an undergraduate at Harvard. Like those of Land and the others, this initial venture was a research enterprise. Formed in 1980, Fuel Tech began by developing fuel saving devices in response to needs exhibited by the oil crisis. Then, as oil prices declined, Haney "converted our firm into a developer of cutting-edge air pollution control technology." In 1987, when Fuel Tech had received more than thirty patents, was employing a staff of two hundred, and had a value of $200 million, its twenty-five year old founder decided to sell the company: "It had become clear that the technology could be marketed more effectively by a firm with more resources than ours." For his next move, the inventor-entrepreneur was determined to "build a world-class firm dedicated to leveraging technological innovation to the natural world's advantage." Within two years he had formed two research firms. One, Energy BioSystems, used revolutionary biotechnological "tools" to "commercialize a microbial system for removing sulphur," a major air pollutant emitted in the processing of fossil fuels. The other, Molten Metal Technology, used a technological base licensed through MIT to develop and commercialize a system to convert municipal, industrial, hazardous, and nuclear waste to useable industrial products. By 1994 both companies were still research enterprises, but their projects were as promising as those technologies that Landau, Land, and Moore had developed before building the enterprises needed to manufacture and market their products.

Haney's goals, like those of his predecessors, are ambitious. As a result of his learning years as a researcher, he has become fully aware of the nation's intellectual resources (its universities, technical libraries, government agencies, and industrial associations) and its financial resources (capital markets, flexible financial instruments, and strategic interfirm alliances). His goals and awareness of these resources bode well for the continuation into the twenty-first century of the dynamic entrepreneurial activities that so benefited the nation in the twentieth by transforming inventions into products made and marketed globally.

THE SECOND-GENERATION ENTREPRENEURS

As might be expected, the careers of the second-generation entrepreneurs, the four who took over enterprises built by their fathers, differed from the first, the six who created their own firms. These four were CEOs, not inventors. Their achievements were of a more corporate than individual nature. Nevertheless, they were entrepreneurial corporate managers. The success of their enterprises lay in the abilities of their companies to move into new lines of business and to adopt new technological processes into their operations. Of the second group two were in manufacturing. The other two were in service industries—one in construction and the other in finance. The differences between the two generations are clearly reflected in the tone and content of their essays.

Thomas D. Cabot, by far the oldest of the contributors to this volume, attended MIT and Harvard before he, at the age of twenty-five, and his brother James took over in 1922 the company their father had formed in 1882 to produce carbon black, the basic ingredient used in ink and automobile tires. Tom and Jim Cabot (and, after Jim's death in 1930, Edward Billings) transformed their well-established, relatively small, personally managed enterprise into a large, modern industrial corporation. They did so by riding the booming tire market that accompanied the massive expansion of automobile production in the 1920s. They moved their production from West Virginia to Texas, where the opening of new oil fields provided the cheapest source of natural gas, their basic raw material. Between 1925 and 1930 the company, using its own engineering force, built nine plants. Employing the most advanced production technology, it so effectively exploited scale economies that its capital and operating costs were less than 60 percent of those of its competitors. To insure a steady flow of raw materials into these plants, the company found its own supply of natural gas and began drilling for oil and gas. At the same time, it built a national and global sales and distribution organization.

With the onslaught of the Great Depression of the 1930s and the collapse of the automobile manufacturing industry, carbon black sales plummeted. In 1940 sales were still only $7 million, of which carbon black provided a little more than a third. The rest was from natural gas, oil drilling, and drilling equipment produced by its engineering organization.

World War II transformed the company and its industry, as it did so many US companies and industries. The Cabot Corporation played an important role in the wartime crash program to produce synthetic rubber, which was crucial after the Japanese occupied major regions that produced natural rubber. That program, in turn, force-fed the development of the new polymer/petrochemical technologies. The Cabot Corporation designed, supervised the construction of, and operated one of the very first plants based on an "oil feedstock, first used as a supplement to natural gas in the furnace process, [which] became the primary raw material that was used," and one that "launched a new era for Cabot as an international company."

Just as the history of the Cabot Corporation provides a model for the successful strategy of corporate growth through vertical integration, which characterized the prewar era, so its growth in the postwar history provides a model of a successful implementation of a standard postwar strategy of growth through geographic and product diversification. The Cabot Corporation, like the more successful US enterprises, continued to enhance the technology of its long established businesses and at the same time moved into new markets based on its existing organizational capabilities as it expanded operations globally.

The core of its growth remained carbon black and natural gas. The company continued to improve its carbon black products and the processes of their production as it moved into new geographical markets. By the 1980s it was the world's largest producer of carbon black, twice as large as its nearest competitor. It transformed its natural gas business by developing a new process to liquefy the gas. Cabot then became a leading transporter as well as producer of natural gas, carrying liquefied gas from distant oil and gas fields to the world's industrial and urban centers. By 1971 it was already bringing huge tankers every few weeks from Algiers to storage and distribution facilities in Boston Harbor. It continued to carry on its oil and drilling network business until the early 1990s when, given the vicissitudes of that industry, its subsidiary was spun off to its stockholders.

Cabot's diversification into other product lines had its beginnings during the Depression when it shifted the focus of its plant-building engineering unit to manufacturing and selling equipment used in drilling for and producing crude oil. During the war it concentrated on building heavy ordnance. It continued to do so until the 1960s, when missiles replaced "big guns." Then it began producing steel casings, but with little success. In 1963, in order to use its metallurgical capabilities, the company moved into the production of titanium, but again with little success. So that unit was sold off to New Jersey Zinc in 1972. The company undertook one more effort in metallurgical products with the acquisition of Union Carbide's Stellite Division, makers of cobalt alloys. This division developed new specialty metal products, but most of the business failed to grow, and most of its activities were sold off during the difficult years of the mid-1980s. The only successful product that was developed was tantalum powder, and Cabot became the world's largest producer. But clearly Cabot's metallurgical capabilities were not strong enough to continue with product diversification.

On the other hand, the capabilities discovered in developing and improving its primary lines remained the major source for Cabot's successful product diversification. One of the first new products, licensed in the 1950s, employed production processes very similar to that of carbon black to produce a specialty chemical, fumed silicon, used in industrial processes for thickening, dispersing, flow control, and reinforcement. Cabot is now its largest producer in the United States and second largest in the world. Another was a line of thermoplastic concentrates (black and white master batches) that were soon produced in plants in Europe and the Far East as well as the United States. The most recent of its new product lines, hearing protection devices, came out of the company's R&D unit. Its commercial success, in turn, led to a line of eye as well as hearing devices. This growth through diversification rested on strong, firm-specific, learned technical and managerial capabilities that helped to make the Cabot Corporation by the 1980s an enterprise with some $1.5 billion of revenue and provided it with a technological base for maintaining its continuing growth and profitability.

The career of Robert Galvin, the other second-generation entrepreneur in manufacturing, parallels that of Tom Cabot, but in a younger company and a newer and faster-moving technology. His father, Paul, formed Motorola in 1928 to produce a specialized product, car radios, for the rapidly growing motor vehicle market. During the Depression the company produced home radios; during World War II it grew rapidly by making some fifty thousand walkie-talkies, two-way radios, and radio devices for military use.

After returning to the production of civilian goods, it reduced its dependence on RCA for radio tubes by producing in 1952 transistors on license from AT&T, first for its own internal use and then for outside customers.

By the time Robert, at the age of thirty-six, took command of Motorola after the death of his father in 1959, he had attended Notre Dame, served in the Army Signal Corps in World War II, and worked for Motorola. Under his command, the company quickly became a leader in the new semiconductor industry, making transistors and then integrated circuits. By 1963 it was the second largest in the United States with 10 percent of the market, behind Texas Instruments (a geodetic instrument company established in 1931), which had 18 percent, and just ahead of Fairchild Semiconductor (Moore's company) with 9 percent. Within the next decade, Motorola and Texas Instruments had become the world's largest producers of semiconductors for the market. (Firms like IBM and AT&T only produced for their own use.) Both built global organizations. By 1970 Motorola had plants operating in Britain, Germany, Latin America, and East Asia. By the late 1970s, according to one estimate, it had a 33 percent share of the US market, just behind Texas Instruments but far ahead of Fairchild's 8 percent and Intel's 6 percent.

Motorola, unlike Intel and Fairchild, produced a broader product line than just semiconductors. During the 1960s Robert Galvin expanded the company's radio and television business and moved into telecommunication equipment. When the large Japanese consumer electronic companies began to mass produce low-cost products, Motorola sold off its radio/television division to Matsushita in 1974. It then, through both internal investment and acquisition, expanded its telecommunications business, particularly modems for local telecommunications and marketing systems. Although Motorola suffered when the Japanese onslaught in memory chips forced it to shut down its DRAM plants, and although it did not have Intel's lucrative PC franchise in microprocessors, it suffered less than other producers precisely because it maintained and expanded a broader line of products. It enlarged its paging equipment output, quickly began to produce cellular telephones in volume, and continued to maintain its strength in microcontrollers in addition to making microprocessors for Apple, Hewlett-Packard, and then IBM. By 1990, when Robert Galvin retired, Motorola enjoyed sales of $9.6 billion; 28 percent of the sales came from semiconductors, 34 percent from communication equipment, 20 percent from general communication systems, and 18 percent from government electronics and other products.

Robert Galvin's Motorola maintained technological leadership in electronics, as Cabot did in chemicals, through continued enhancement of existing products and processes and the commercialization of new ones. This basic strategy of long-term growth by improving existing businesses and developing new ones based on the company's learned specialized capabilities was similar to what Stephen D. Bechtel, Jr. used in engineering-construction and Edward C. Johnson 3d in finance when they took over their enterprises from their fathers.

When Stephen Bechtel became president of Bechtel Group, Inc. in 1960 at the age of thirty-five, his company's revenues were $460 million; when he retired in 1990, they stood at $8.9 billion. Stephen received his engineering training at Purdue while serving in the Marine Corps Reserve during World War II. He then earned an MBA degree before entering the company, where he soon moved up the ladder to become the manager of the Pipeline Division and then the Treasurer of the Corporation. During his tenure as President and Chairman, the company expanded its existing activities in the construction of pipelines and of oil refining and chemical plants, particularly those producing the new polymer/petrochemicals. At the same time it moved into related businesses, such as building nuclear power plants and those to liquefy natural gas; constructing airports, shipyards, and other transportation works; and erecting military and space facilities for government programs and for environmental protection and cleanup. In carrying out the last of these, the company developed an expertise in "the technical and practical aspects of air, ground, water, and sound pollution." In all these lines of business the company incorporated new electronic information and communication technologies and used the new materials developed by the new chemical and metallurgical processes. Over these same years Bechtel became a truly global enterprise. In Stephen Bechtel's words: "One of Bechtel's great strengths has been its geographical and industry diversification."

But continuing success was in no way assured. By the early 1980s the combination of sharply fluctuating oil prices, recession, inflation, and intensified competition became an even greater challenge to Bechtel than it was to Landau, Moore, Hatsopoulos, and Galvin. The completion of existing contracts delayed the full impact of the drastic downturn in the engineering construction business until 1983. Between 1983 and 1987 Bechtel's revenues fell from $14.3 billion to $4.5 billion, and its work force dropped from 44,500 to 17,400. However, Stephen Bechtel successfully met the challenge of downsizing. In 1990, Bechtel, with its revenues of $8.9 billion and a staff of 21,400, remained one of the world's largest and best enterprises in its business.

Ned Johnson's father, Edward, a lawyer who had formed Fidelity Investments in 1946, pioneered and became a leader in the development of the still new mutual fund business, that is, funds that held extensive portfolios of securities. During the boom years of the 1960s, the company prospered, especially with its innovative growth funds. But when the son became its CEO in 1972, Fidelity was reeling from the collapse of the bull market of the 1960s. In the two years between 1972 and 1974, its financial assets fell from $5.5 billion to less than $1.7 billion.

The younger Johnson, a 1953 Harvard graduate who had worked for the company as a financial analyst and a fund manager during the 1960s, met the challenge with a twofold strategy. In addition to creating new products and concentrating on better performance, Fidelity moved forward into marketing and distribution and backwards into critical ancillary activities (much as Cabot and other manufacturing firms had done as they built their integrated enterprises earlier in the century). In 1974 Fidelity introduced its first money market fund, Fidelity Daily Income Trust, that allowed customers to cash checks on their accounts. In marketing and distribution, Fidelity was the first "to sell our funds directly to individual investors through direct response advertising and a toll-free telephone line," while it also continued to sell through brokers. To further enhance its distribution, it entered the discount brokerage business in 1979.

Johnson then began "to bring the shareholder processing operations inside." Fidelity became responsible for the transferring of funds and securities and all bookkeeping and accounting that had previously belonged to banks and other institutions. These activities were made increasingly more productive by careful integration of computers and other information and communication technologies into their operations. Fidelity also continued to introduce new, more specialized funds, including those that invested primarily in specific individual industries, in emerging markets, in municipal bonds and in high-yield—or "junk"— bonds.

With a variety of innovative mutual funds available to the market, as well as a solid organizational infrastructure built on a strong technological foundation, Fidelity was well prepared to exploit the opportunities of the bull market that began to flourish after 1982. The number of mutual funds the company offered increased almost exponentially, and more than a hundred different funds were available from the company midway through the decade. By the end of 1995, Fidelity was the world's largest mutual fund company, with over two hundred thirty funds and more that $400 billion in managed assets. The past ten years also saw the company expand its business sights well beyond its core mutual fund and brokerage operations. Fidelity Capital, for example, was organized to provide business expertise and financing for Fidelity-owned start-up companies. As these companies grew in size, they became autonomous subsidiaries partially owned by their staff, much in the manner of Hatsopoulos's technological "spin-outs." Key elements of Fidelity's business were created, nurtured and built in such a manner, creating new and entirely different lines of business even as its financial services operations continued to flourish.

THE IMPLICATIONS OF THESE ENTREPRENEURIAL ACHIEVEMENTS

This brief review of the companies these entrepreneurs created or expanded tells much about the role of the inventor-entrepreneur and the entrepreneurial manager in what economists have termed "technological advance," a basic source of a nation's economic growth. The first generation of corporate builders just described all had a commitment to and a natural aptitude for invention. All received a high level of technical education—both engineering and scientific. MIT clearly was an incubator for this breed of entrepreneurs. The commercializing of their inventions here and abroad called for the ability to create a business enterprise. All their enterprises began as research organizations whose revenues came from selling licenses and doing contract work for corporations, government agencies, and other institutions. The critical challenge was building their own corporations to produce, distribute, and sell their products in national and increasingly global markets. All had to become managers, CEOs, or, as in the case of Taplin, turn the management over to a professional CEO. If the processes of production and distribution encouraged the making of standardized products in sufficient volume to assure the cost benefits of scale economies, the resulting enterprise became or was sold to a large global corporation.

For the entrepreneurial CEO of the second generation, the challenge was to maintain and expand what was already a successful large business enterprise. Here neither an inventive instinct nor a MIT type of education played a significant role. Instead the challenge lay in encouraging their teams of managers to enhance the firms' existing capabilities and to seek the new opportunities created by ever-changing markets and technologies that would best use the firm's existing capabilities. For all, the basic strategy became the development of a stream of new products and businesses whose development and commercialization were funded primarily by retained earnings. (Virtually the same strategy was used by Hatsopoulos to further the growth of his first-generation enterprise.)

The financial and competitive success of the enterprises that resulted from the entrepreneurial achievements described here speaks well for the future of American industry. But, except for Taplin, all these entrepreneurs and their enterprises were faced with an unexpected and severe reduction in their stream of revenues. The original title for this collection of essays, suggested by Elkan Blout, was "American Industry at the Crossroads." For these firms, such a crossroads came from the late 1970s through the mid-1980s. Those years were times of trouble for Landau, Land, Moore, Hatsopoulos, Galvin, Bechtel, and, to a lesser extent, Cabot. The causes were multifold: the second oil shock; stagflation (inflation without growth); the high cost of capital (about which Landau and Hatsopoulos write perceptively); the crowded markets, particularly as Europeans and the Japanese caught up to and in some cases surpassed US firms in terms of technological productivity and capabilities; and the increasing costs of integrating the new technologies into existing operating procedures and processes.

For many other US firms these years were even more difficult than they were for those described here. In the 1960s and 1970s many industrial enterprises had grown by diversification and through merger and acquisition into businesses in which their organization's own learned capabilities gave them little or no competitive advantage. For these firms the late 1970s and the 1980s were years of downsizing and reshaping their product lines so that they reflected the firm's basic core competencies. By the late 1980s this restructuring had been effectively carried out so that most US firms and industries were again competitive in world markets.

This was particularly true in the high-tech industries in which the manufacturing companies mentioned here were concentrated. In electronics and chemicals as well as pharmaceuticals and aerospace, US firms are powerful competitors. Although American companies lost out in consumer electronics (in good part due to the gross mismanagement at RCA), they retained their strength in telecommunications and are world leaders in the cutting-edge technologies of data processing, including both hardware and software based on the microprocessor. In chemicals and pharmaceuticals they are in the vanguard (and particularly in the new biogenetic technologies), as they are in aerospace. Many of the leaders in these fields have larger revenues from abroad than from home. The Japanese are not yet significant players globally in either chemicals or pharmaceuticals, nor in aerospace. The Europeans maintain their strength in chemicals and pharmaceuticals and, to a much lesser degree, in aerospace. But they have lost out almost completely in consumer electronics and in computer hardware and software operating systems—the most dynamic of the postwar technologies.

Angus Maddison, the foremost authority on economic growth, has rightly argued that "technical progress is the most essential characteristic of economic growth. If there had been no technical progress, the whole process of accumulation would have been much more modest."[1] At the heart of technical progress is technological advance, that is, the innovation and commercialization of new technologies. Large firms, like those of the second generation described here, are central to the process. Indeed, the successful challengers to US industries in global markets have primarily been comparable large European and East Asian industrial companies. But in Europe and East Asia, inventor-entrepreneurs who can build the enterprises necessary to commercialize products in global markets remain a rarity. And as the youngest author, William Haney, so eloquently points out, the United States has unparalleled resources—educational, financial, informational, and institutional, in both the public and private sectors—with which to nurture that critical ingredient of technological advance: the first-generation enterprise that has brought many new technologies on stream.

ENDNOTE

[1]Angus Maddison, "Explaining the Economic Performance of Nations, 1820–1989," in William J. Baumol, Richard Nelson, and Edward Wolff, eds. *Convergence of Productivity: Cross-National Studies and Historical Evidence* (New York: Oxford University Press, 1994), 53.

John F. Taplin

The Education of an Inventor

I AM ALWAYS AMUSED WHEN I SEE an advertisement that announces a new, "all-natural" product. The fact is that everything comes from nature, and the only thing that is new is uncovering nature's secrets. An inventor may modify something and reshuffle its components, he may analyze it and predict how it will behave, but he always starts with natural things that are ruled by the laws of nature. And he almost always builds on discoveries that other people have made, often rearranging or combining their findings.

Essentially everything I have done in my working life has been in accordance with that basic principle. You find out what people want to do better; you analyze what they are doing and why they are not getting the results they want. Then you use elements of prior discoveries in conjunction with what is available from nature until you have an invention that will do what you want. That, in a nutshell, is *invention*. I came to this understanding through both my upbringing and my education.

Born in 1913, I was the second of five children. My father had studied classics at Amherst College. He then became a social worker in Boston, setting up and running settlement houses to help the Irish who were arriving in great numbers. His income as a social worker was quite limited, so my siblings and I attended public school before entering college. We lived in a comparatively unpretentious house in Wellesley, near Wellesley College. My mother taught philosophy at the college to augment our family income.

John F. Taplin is the Managing Director of Harvard Medical School's Funds for Discovery Program.

My older brother was interested in the classics and later became a professor of classics and English literature. My younger brother loved music and became an organist and choir director. My main interest, however, was neither in the classics nor in music. I loved to make things, to do things with my hands.

My mother's brother, Edward Oakman Hood, after graduating from Lawrence Scientific Engineering College in Cambridge (which later became the School of Applied Science at Harvard University), lived with my grandparents in Wellesley Hills. He was fascinated by electronic communications (which was then in its infancy), as well as by other types of communications, in electrically energized fluorescent tubes and in photography. His wide interests related to the exploding "scientific curiosities" of the time—things he could physically demonstrate, handle, and feel. His father, a Congregational minister, did not encourage him to become a scientist, and so he joined the Roger Babson Company and advised clients on ways to invest their assets.

My uncle soon found that analyzing business reports was not compatible with his personal scientific interests. He therefore accepted a position as a Sunday School teacher in the Wellesley Hills Congregational Church and taught his own course, "Miracles Abound." I spent every Sunday morning for ten years watching my uncle demonstrate his "scientific curiosities." My uncle was not only able to please his father by teaching a Sunday School class but the class allowed him to pursue his scientific interests.

He entertained us with electric "magic" by stringing electrical wires around the Sunday School's hallways so that we could talk to each other throughout the building. My uncle would assemble glowing tubes of neon gas, and then bend them with a gas torch until they spelled our names. He rationalized that these "Wonders of the Lord" demonstrated what a clever fellow the Good Lord was in providing us with all these "natural building materials."

My uncle was an expert photographer, and he enlisted me as his assistant. In those days ready-made film was very expensive. So my uncle would cut a piece of clear glass to fit his film-holder, then dip the glass in light-sensitive emulsion, and finally expose the film in his big box camera. When we took indoor photos, I prepared the photographer's flash light by sprinkling magnesium powder into a hand-held trough. When he gave the signal, I would strike a sparker and the magnesium would give off a brilliant flash, producing a lot of smoke. We would then go into the darkroom and develop the exposed plate. It all seemed like magic to me.

My first exposure to a fully automatic machine was a music box that my grandparents had purchased in Switzerland. The mechanics of it absolutely enthralled me. There were gears to change the speed and a propeller that acted as a governor to control the speed of the rotating music cylinder. One afternoon, when my grandparents were napping, I partially disassembled the music box to better understand how it worked, though I was careful not to dismantle it to the point where I could not get it back together again. For my Christmas present the next year, they gave me a smaller Swiss music box.

The following Christmas my uncle gave me a Meccano Set, produced in England, that was designed for children who were seven or eight years old. It was the forerunner of the Erector Set and was the perfect construction toy for a boy who liked to assemble mechanisms to perform specific tasks. It contained girders, gears, nuts, and bolts so that one could construct things such as wheelbarrows, carts, and very simple machines and toys. The manufacturer included with the set several booklets that listed all of the parts for their other eight construction sets, including drawings of the various things that could be constructed. The following year I received a more advanced set, a number two set, with more parts so that I could build bigger and more complicated mechanisms. Each year for the next six years my uncle gave me the next larger construction set. As Christmas day approached, I started designing more complicated mechanisms in anticipation of receiving the new Meccano Set from my uncle.

Meccano published a monthly magazine, distributed worldwide, that showed some of the mechanisms and vehicles that were not listed in their instruction booklets. My ambition was always to be an inventor and my uncle constantly challenged me to invent new things. The year that I received the number four set, which came with a miniature electric motor—the first one I had ever seen—and a transformer to run the motor, I designed and made a trolley car that actually ran on tracks. My uncle photographed me standing beside my trolley car and the photograph was sent to the Meccano Company in England along with my plans for putting it together.

Two months later, my name and a description of the new Taplin-Meccano Trolley Car were listed in their monthly magazine.

As an adolescent I constructed many things, some useful, some just for fun. I wired the front doorbell so that I could ring it from my room upstairs. I designed an automatic "window-closer" for the windows in my room, so that no one would need to get up to close them when it started to rain. My best invention was the water-heating coil that I constructed and placed in the chimney flue of the furnace. This homemade heater made hot water automatically, at no cost to my family, from the exhaust hot air that otherwise would have gone up the chimney.

My uncle took me to visit a large paper coating factory in Dorchester that used one large steam engine to drive all the factory's machinery. The giant steam engine was equipped with an 8 foot drive wheel, and belts attached to the drive wheel ran throughout the factory to supply power wherever it was needed. The plant's operations engineer explained the mechanisms that controlled the action of the engine. I spent several days visiting the paper factory; each time I was amazed to see the tremendous power that came from just one coal-fired boiler supplying steam to drive a single steam engine.

Every Sunday, after returning from church, we would have a very formal meal. The dessert was always ice cream, and my brother and I were responsible for hand cranking the ice cream freezer. Turning the freezer crank for half an hour was a task we tried to avoid whenever possible. One Sunday, when I was left alone with the job of turning the crank, I decided that there had to be a better way to freeze ice cream.

My grandparents had given my mother a new Easy Washing Machine, the first electric washing machine that Wellesley had ever seen. As I examined it I noticed that there was an electric motor underneath the machine, and I wondered if this motor could be used to crank the freezer. The motor was secured with four bolts, and luckily it had a splined shaft so that it could be removed from the washing machine without difficulty. When the motor was connected to the house current, it ran beautifully except that its drive shaft turned much too fast. Somehow, the speed had to be reduced. I remembered that in mowing the lawn the mower's wheels ran much slower than the rotating cutting blades.

Why not use the lawn mower to reduce the speed of the washing machine's motor? This was accomplished by using a series of belts and pulleys from the washing machine's motor shaft to the mower's cutting blade shaft and a drive belt from the mower's wheels to the crankshaft of the freezer. When I connected the motor to the house current, the freezer's shaft turned at the correct speed of sixty complete turns each minute!

I thought it would be best not to show my parents what I had invented, but I was so proud of my new labor-saving machine that I had to show my brothers. Later that day my mother noticed that the motor was missing from the washing machine, and she was horrified. This new washing machine was the most important labor-saving device in our house. But when I showed her how quickly I could remove the motor from the freezer and reinstall it on the washing machine, she allowed us to borrow the motor to make ice cream whenever we needed it. That was the first automatic home ice cream freezing machine in Wellesley and maybe in the whole United States.

It was my uncle who showed me that new ideas first consist of scientific curiosities and that they later can be produced in quantity and marketed for the benefit of the public. He encouraged me to continue to develop unusual labor- and energy-saving experimental devices by inviting me to demonstrate my wonderful new inventions to neighbors and friends. My parents also showed interest in these scientific curiosities even though they did not understand their mechanisms.

In high school I was most fortunate to be taught by Jasper Moulton, an outstanding mathematics teacher. He introduced me to algebra and spoke persuasively about the inherent beauty of mathematical logic. I was given special instruction in plane, solid, and analytic geometry, and even elementary calculus. Mr. Moulton posed unusual geometrical problems for me to solve as my "special homework." The extra time that Mr. Moulton spent instructing me in analytical reasoning strengthened my resolve to pursue more advanced study in mathematics and logic later in college. Both Mr. Moulton and my uncle encouraged me to attend the Massachusetts Institute of Technology (MIT). My parents were concerned about the tuition expenses, but Mr. Moulton assured them that I would obtain scholarships and student loans to cover the tuition fees.

At that time MIT required applicants to take thirteen separate entrance examinations, covering many of the subjects that I had studied: French, German, English, physics, five different branches of mathematics, etc. All the boys at Wellesley High School who were planning to go to college left high school in their junior year to attend "preparatory schools," such as Exeter and Andover. I was, therefore, the only boy in my class who planned to go to college directly from Wellesley High School. In some ways this was an advantage because the high school gave me special attention and coaching in preparation for my college entrance examinations. MIT provided me with generous resources to support the tuition expenses for all of my undergraduate engineering education. I lived at home and traveled to and from MIT by trolley car.

Professor Norbert Weiner, a famous mathematician at MIT, was my favorite teacher. I learned both theoretical and applied mathematics from him, including new theories that were just being applied to solve stability problems. He taught me how to transform all types of wave forms into simple mathematical expressions and explained how the wave form of the human voice could be represented by a continuous frequency spectrum for transmission over copper wires and recombined at the far end of the transmission line to eliminate distortion.

Another of my instructors was Professor Ernest Guilleman. He constructed complex equations using understandable and logical explanations of current electronic circuits for transmitting voice and data between two distant points with little distortion. We take low distortion transmission for granted now, but at that time Professor Guilleman was contributing to the design of unique electric circuits that made long-distance telephony possible. It was all achieved through the application of his original mathematical expressions.

The most important lesson that I learned at MIT was that one could accomplish almost anything if one was able to express the logic behind it and then apply these logical statements to achieve the desired results. The high level of self-confidence that I developed in understanding how to improve the performance of products or processes by describing their actions in basic logical expressions prepared me for a lifelong career as an inventor. The four years that I spent at MIT were not only filled with exciting experiences, but they taught me how to invent new ways to design and produce less costly and more reliable devices and mechanisms.

After graduation, as a controls engineer for the Foxboro Company, I started to investigate the causes of undesirable, continuous oscillations in controlled systems. For instance, microphones placed too close to loudspeakers produce high-pitched oscillations, the irritating screams we call "feedback." This condition is caused by the excessively high level of acoustic energy from the oscillating loudspeakers "feeding back" into the sensitive microphone. Other kinds of excessive feedback conditions—for instance, in the automatic pilots on ships and airplanes—can develop similar oscillations.

After spending several years investigating feedback systems in industrial processes, a research partner and I were commissioned to write a book describing stability criteria in feedback control systems. McGraw-Hill published this book, entitled *Automatic Feedback Control,* several years after I graduated from MIT. This book was promoted and sold as the standard text for studying the dynamic analytical equations that defined system stability under automatic control.

When I graduated in 1935 the country was in the depth of the Great Depression, and I could not locate a job that would use the skills and engineering knowledge that I had received at MIT. The fact that I did not get an engineering job, however, turned out to be to my advantage. I was always fascinated by how things were made. I remember examining beautiful windup toys produced in Japan that were complex mechanisms and yet were sold for only twenty-five cents. I wondered how the Japanese manufacturers could construct these products and sell them for so little. I learned that the Japanese bought discarded beer cans for very little, shipped them to their factories in Japan, and then processed them through machines, which cut them open and flattened them. These small sheets of metal were fed through automatic punch presses and formed into dies to produce the parts for these Japanese toys. Sure enough, when these toys were taken apart—what is referred to as reverse engineering—one could still see the beer can labels on the insides.

The Foxboro Company produced instruments with somewhat similar production machines. I went to them with the hope that they would hire me as an engineer and toolmaker. They asked about my qualifications. When I told them that I had made mechanisms by hand and that I had an MIT degree, they said that I was overqualified. When I offered to work for twenty dollars a week, however, they hired me as an apprentice toolmaker.

For the first three years after college, I was a toolmaker. The most expert toolmakers were trained in Germany. It was their job to design and make the tools that enabled the company to hire workers at starting wages to load metal stamping machines. The tolerances in making these tools were one tenth of the thickness of newspaper. This was the kind of experience graduate engineers rarely receive: designing and then personally making the actual tools of production.

This hands-on experience is important for corporations that wish to produce and sell quality products at a reasonable price to compete with foreign manufacturers. Engineers should be able to design products that toolmakers can make easily and that will require low maintenance. Toolmakers often have to go back to engineers, who have had little hands-on practical experience, and say, "If we make this as you designed it, it will be expensive. How about designing it this way?" Thus the practical toolmakers are often required to show graduate engineers how to revise their production drawings to produce more reliable products at a lower cost. Working in a tool-design and toolmaking department provides the best training for anyone who wishes to be a highly qualified product design engineer.

When the war broke out, one of my professors at MIT arranged for me to redesign small gasoline engines to drive electric generators in Navy bombers. When this became a full-time job, I left the Foxboro Company and went to work for a company in New Jersey as their engineer in charge of designing new engine speed controls. The engines in question had already been manufactured and were ready for shipment when the military specified that they had to be equipped with speed governors. The generator was designed to produce alternating current when running at a speed of 1,800 revolutions per minute. Excessively high speeds would damage the engine, and speeds that were too low would not generate sufficient electricity.

My assignment was to design a special governor—one that did not require the conventional governor with a drive shaft. I noticed that all of the carburetors in these engines produced a significant amount of low air pressure to operate the engine's throttle. Using this low air pressure to energize the governor, we were able to rotate it without a drive shaft by simply mounting the governor on the outside of the engine's existing flywheel. It was an original design of a new type of vacuum governor. I was awarded basic patents on all classes of high performance vacuum governors. Thousands of these governors were used to control the speed of engines driving electric generators in aircraft during World War II.

About two years before the war ended, an associate working for the government noticed my design of this high performance governor used to control the speed of gasoline engines. After running tests of its performance and noting the simplicity of its mechanism, he and his partner, Dr. Carl Walter, invited me to be the Director of Engineering in their new company—the Fenwal Company. Carl Walter, a graduate of Harvard Medical School, had formed the Fenwal Company in Ashland, Massachusetts. He had observed that blood for transfusions was stored in glass bottles. These glass containers were much too fragile to transport blood during World War II. Dr. Walter started Fenwal not only to develop containers for transporting and handling blood, which would eliminate the problems with glass containers, but also to produce more reliable thermostatic controllers.

I was married with one small child when we moved to a town not far from the Fenwal Company. One of my first assignments at Fenwal was to work with Dr. Walter to develop a rugged plastic bag suitable for storing and transporting human blood. It was an arduous task. The plastic first used was difficult to seal and contained ingredients that leached out, thus contaminating the blood. After many months of laboratory work a plastic blood bag was produced that could stand up to all the specified tests. This original Fenwal blood bag soon became the worldwide standard for transporting and handling blood. Even today the Fenwal blood bag is used by millions in all countries throughout the world.

Next the Fenwal Company began developing a dialysis machine that could be used as a temporary artificial kidney. Dr. Walter asked me to assist him in designing a parallel plate artificial kidney unit. This was the first practical parallel plate dialysis unit made, and Fenwal patented it. We then turned our attention to the design of artificial aortic check valves for human use.

When the war ended, Fenwal had to downsize as a consequence of a substantial reduction in government contracts. In view of this new situation I decided to leave Fenwal and start my own company. I had put aside savings for this event, but unfortunately not enough to buy new equipment. I was advised to procure the equipment, instruments, and tools I needed by bidding on these items at auctions. These auction purchases enabled us to obtain machine tools—in first-class shape—that we could program to produce high quality parts. By buying all of our needed tools and equipment at auctions, we saved two-thirds of the start-up cost for new machinery and equipment.

Originally this new company—the Kendall Controls Company—was located in Kendall Square in Cambridge, Massachusetts. In less than one year we moved to a substantially larger building in Waltham. Several factory workers, two machinists, and a product engineer were hired to work in our new shop. I served as sales manager, bookkeeper, and new product designer.

The Foxboro Company—where I had investigated the causes of undesirable system instability after graduating from MIT—manufactured a wide range of pressure regulators. They were expensive and unable to control the pressure within the limits required for many critical measuring applications. A market study reported that there were a substantial number of companies that had immediate applications for high-performance, superior-quality pressure regulators. This open market report supported my interest in designing, producing, and selling superaccurate regulators.

These new superaccurate regulators that we developed and patented were not only smaller than the Foxboro regulators but were substantially more accurate and easier to fabricate. They were also designed to be made with die castings and punch-pressed parts using production processes similar to those used by the Japanese in the production of their low-cost toys. These easy-to-make regulators allowed us to put new employees to work with minimal training. We found the production costs were so low that we could produce the highest quality products at a cost of only 20 percent of their selling price. This profit margin was accumulated to support the company's future expansion.

When the company first started, and we were seeking new customers, I took our first samples of the Kendall high-performance regulators to the Chief Engineer of the Foxboro Company. "I have developed a pressure regulator that has exceptionally high accuracy and excellent performance," I said, "and I am offering our regulators to you at half the sales price of your present Foxboro regulators." I left some samples with the Chief Engineer in the hope that his engineering department would run performance tests.

The Foxboro Chief Engineer called us several weeks later and asked for an exclusive worldwide license to manufacture and sell the Kendall regulator. After reviewing this offer with our sales group, we decided not to give them these rights. Foxboro instead bought large quantities of Kendall regulators directly from us and sold them through their worldwide sales office. To fulfill these large orders, we increased our manufacturing facilities and hired additional factory workers. The Kendall Company was launched with the expectation of a rapid increase in the sales of a profitable line of superior regulators.

At that time there were ten other companies making and marketing pressure regulators. We commissioned a marketing consultant to find sales representatives who could promote the Kendall pressure regulators throughout the country. To introduce our new representatives and products to prospective customers, I went with our salesmen as we demonstrated the accuracy of our Kendall regulators. Many of the companies we visited subsequently purchased Kendall regulators in large quantities.

The Sheffield Air Gauge Company produced thousands of air-operated instruments that controlled critical parts used in truck and automobile engines. We had learned that because of the pressure changes in air distribution conduits, Sheffield could not provide its customers with instruments that met their required level of performance. Sheffield was not able to make or buy a regulator that met its customers' specifications.

Our field representatives and I visited the Sheffield Company to demonstrate the accuracy of our Kendall regulators. The engineers determined that it passed their customers' most stringent tests, and Sheffield insisted on an exclusive worldwide license to make and sell our regulators. This time I contacted a law firm and provided their lawyers with estimates stating that Sheffield sold over 50 percent of all air gauges, and that their customers were starting to demand a substantial increase in the ability of the Sheffield instruments to meet higher performance standards. This information enabled us to negotiate a license arrangement specifying that our company's regulator would be the only one used, that our company alone would produce all their regulators, and that we could price our regulators to include a 25 percent profit margin. Thereafter we made and sold many thousands of regulators to Sheffield for use in all their air gauge instruments.

We next built a larger plant to manufacture these regulators and other related products. Our annual sales approached $25 million. Our expanding company never needed to borrow funds because our retained earnings were always in excess of our requirements for growth. This method of financing our growth amazed both Harvard Business School and MIT professors. They offered to review our operating expenses and pricing procedures in order to prepare a case study of these "unusual company practices." When they learned that all we did was create original products that companies want, protect them by US and foreign patents, and then price the products with a sufficient profit margin to self-finance future expansion expenses, the case study projects were dropped.

Our home sales department noticed that a number of our regulator customers had begun to order hundreds of extra rubber and fabric diaphragms that we made as an internal part of our pressure regulators. I had designed and patented this new fabric and rubber seal for sealing many types of pistons. We soon learned that the companies that were ordering these diaphragms had improved the performance of their own instruments and solved problems that they were having with their own sealing devices by installing our diaphragm seals.

The sales department's market study suggested that our own recently patented rolling seal would soon increase our sales levels, surpassing that of our regulators, if we produced and marketed a wide range of rolling seals. This new product undertaking would obviously require a new, larger manufacturing plant with special machinery to produce both regulators and "rolling seals" in one factory.

At that time I had been thinking about selling our oldest line of regulators—ones with limited residual patent life. This sale would provide over $700,000 to finance a manufacturing plant with many specially-designed production machines. The Fairchild Corporation was searching at that time for additional industrial space. We sold our plant and with the proceeds built our new plant on twelve acres of land on Route 128 in Burlington.

This plant was equipped to produce our new line of pressure regulators in addition to a wide range of our patented rolling diaphragm seals. We were confident that this patented rolling seal would find thousands of new customers searching for an ideal seal—one that would eliminate the leakage, friction, and lubrication problems associated with all the piston seals available at that time. Our rolling diaphragm seal eliminated any possible leakage of gas or fluids because the seal filled the entire space between the piston and the cylinder wall. In addition, our seal required no lubrication and, because it rolled, was frictionless.

The fast-growing sales of our new line of rolling diaphragm seals prompted our expanded department to suggest a new name for our company. We decided on "Bellofram" since our product combined the properties of both diaphragms and bellows. The Bellofram Corporation designed and printed a product catalogue with descriptions of all the standard sizes and pressure ranges—all the way from very light pressures up to five hundred pounds per square inch—and charts showing the stroke ranges. Very substantial new commercial sales developed in a few months, especially in the automobile industry. Automotive engineers were interested in our Bellofram seals because they were inexpensive and required no lubrication.

Our company grew rapidly by adding new Bellofram sales offices in the United States and by licensing companies to make and sell Bellofram seals throughout the industrialized world. Our foreign patents enabled us to license companies throughout England, Europe, Australia, and Asia. Again, because we sold our seals at prices that provided margins to support our growth, we were able to use the company's retained earnings to pay for all our domestic and foreign expansion expenses. In 1965, as soon as the company was operating successfully, the complete responsibility for the overall management was placed under the authority of an experienced professional president and chief executive officer. This change in management enabled me to concentrate on selecting, designing, and working on new inventions. I was also able to use a large percentage of my time to design new products that passed our "new product eligibility criteria."

In the late 1960s and early 1970s I consulted with the New York Blood Center in New York City. They were searching for a new way to quickly and accurately identify blood types stored in their blood bank. I collaborated with their scientists in designing a system that would enable them to label their blood bags using pressure-sensitive labels with unique bar codes, printed with magnetizable ink. These labels could be read both by humans and by automatic reading machines. These automatic readouts were accurate even if the label surfaces were dirty or bloody.

When the patents were issued for the automatic reading machine, I established the Taplin Reading Machine Corporation (the name was later changed to the Context Corporation) in Burlington to produce and sell machines that read typed and printed data. After three years, when the company's annual sales were several million dollars, the Burroughs Company offered to buy the Context Corporation. They wanted to use the Taplin Reading Machine to automatically identify and sort checks cashed by bank customers. The purchase price offered by Burroughs was in excess of the appraised value of the company, and at that time our reading machine was fully developed and in use by many customers. The enjoyment that I had experienced inventing and organizing a company to produce and market the Reading Machine was coming to an end. Given these circumstances, I decided to accept Burroughs's most generous offer.

Not long after Burroughs bought the Context Corporation, it seemed best to consider selling the Bellofram Corporation. The untapped future sales potential of the unique, patented products in Bellofram's inventory would best be realized by merging with a firm that had a global sales organization producing and marketing somewhat similar industrial products.

At that time Bellofram's total annual sales were $35 million with a handsome pretax profit. Kidder, Peabody, in New York, was commissioned to find a buyer and negotiate the terms of the purchase and sales agreement. This entire transaction was completed by Kidder within six months.

After these two companies were sold, I thought at first that I would retire, as was the conventional practice. But I was persuaded by my close friend Dr. Carl Walter—the founder and president of Fenwal—to first review his analysis of the expected life span of retirees. The facts and figures in his report clearly proved that an individual retiring after many years of intensive work generally has a shortened life span. These retired individuals found it difficult to adjust to their new living conditions. Since they no longer had a definite daily schedule and were not expected to arise in the morning at any specific time, many felt that it was this dramatic change of pace that directly affected their longevity.

Dr. Walter introduced me to the Dean of Harvard Medical School, Dean Tosteson, who suggested that Harvard Medical School might be able to apply my experiences and knowledge in translating research scientists' ideas into lifesaving biotechnological products. He appointed me as a consultant in the Medical School's Office of Technology Licensing. The school's internal resources to support its scientists' requests for funding their research projects were quite limited. To provide the school with additional resources I established a new program—The Funds for Discovery Program—by transferring the assets of a previous program funded by the Taplin family. The Dean of Harvard Medical School appointed me the Managing Director of this new program.

The Funds for Discovery Program finances a number of seminars run by development-minded individuals who have experience both with academic scientists and with the industrial procedures pertaining to licensing agreements with academic institutions. These seminars acquaint the faculty with the availability of the program's funds to support developmental research projects. Our first seminar described ideas in terms of intellectual property, stressing the fact that real estate is protected by legal documents called deeds whereas intellectual property is protected by patents.

A substantial number of corporations in America are bequeathed to the heirs of the company's founders. These heirs frequently receive a large percentage of the operating profit, and these assets are quite often not reinvested in their own firms. As a consequence, many of these companies lack the financial strength to support product improvement or new product research and development programs, or to install modern and more efficient production machinery and equipment.

Many firms in Asia and in Europe, by contrast, provide the finances not only to support their product research and development departments, but also to upgrade their manufacturing plants with modern labor-saving equipment. Therefore, as companies in these foreign countries continue to renew themselves it will be important for our production facilities to expand their research and development programs and to install state-of-the-art manufacturing equipment. The goods and services produced by these "modernized" companies would thereby enable the United States to maintain its position as the world's strongest economy.

Several months ago, a recent graduate from Cal Tech was shown on television walking up and down the highway in Pasadena holding a large sign that said, "Somebody give me a job. I have a Ph.D. but no job." I thought it odd that an individual, after having spent thirty years of his life learning what he could do to help himself and society, could do nothing for himself or anyone else unless and until someone gave him a job.

I said earlier that I was fortunate to receive an education at MIT during the depths of the Depression. Since I was not able to find a job in the engineering profession when I graduated that would use my technical knowledge, I took a temporary job as an apprentice toolmaker. Within three years I had assembled sufficient tools and equipment to set up a small fabrication and test facility. This laboratory was expanded and new technical personnel were hired to accommodate the rapid growth in developing unique, superior products.

If I have any regret about my life, it is that I did not spend more time with younger people, showing them how to look at something and ask, "How can we make that work better?" It is still true today that most things do not do what we would like them to do and do not work as efficiently as possible. All one needs to do to be an inventor is be aware of the many opportunities that exist for us to "make things work better." The challenges are there, just as they have always been, and so are the theories, the materials, and the processes. If this essay inspires one person to meet one of those challenges, I will have fewer regrets.

In closing, I would like the reader to consider Albert Szent-Györgi's simple and yet comprehensive definition of discovery: "Discovery is seeing what everybody else has seen, and thinking what nobody else has thought."

Ralph Landau

Entrepreneurs, Managers, and the Importance of Finance

T HE WORLDS OF TOP MANAGEMENT and of technologists are often at loggerheads—neither seems to understand the other's perspective. Even when the CEO is an experienced technologist, which is frequently the case in young, entrepreneurial, technically-based firms, he or she must of necessity become increasingly immersed in the business and financial problems of the organization; this leaves little time to keep up with technological developments. Because I was a technologist who started as an entrepreneur and then became a CEO, perhaps I can shed some light on both viewpoints. For years technology drove our business; the 1950s and 1960s were a time of rapid world and industry growth, and superior technology was indeed in demand. However, as the external environment changed in the 1970s, I was confronted with a new set of problems, mostly financial, that converted me more fully into a CEO. In an era of worldwide capital constraints, the lessons that I learned then and thereafter as an economist at Stanford University may be helpful to the technical community of today in understanding some of the key problems faced by technology-based American companies and their managements.

In 1946 I cofounded Scientific Design Company, Inc. (it later became Halcon International and Halcon SD Group). Since the venture capital markets did not yet exist and banks avoided lending to small technology-based companies, there was no outside

Ralph Landau is Consulting Professor of Economics and Co-Director of the Program on Technology and Economic Growth at the Center for Economic Policy Research at Stanford University.

funding during our first twenty years. We grew from within, managing technology for our strategic growth. Some may describe the period from 1946 to 1966 as the golden years of America's postwar dominance of the international economy.

In past writings I have defined an entrepreneur as one who brings together people, money, markets, production facilities, and knowledge to create a profitable commercial enterprise. As cofounder and ultimately chief executive of one of the most successful chemical engineering design firms in the world and subsequently the cofounder and comanager of a large chemical manufacturing firm, I helped develop and commercialize nearly a dozen important, innovative chemical products and processes that are in use today—a remarkable record for any company, and especially for a small entrepreneurial company operating in an intensely competitive field. In this way, I played a significant role in the postwar growth of one of our most important industries.

Nothing in my chemical engineering training prepared me to be an entrepreneur or a manager. Awarded in 1941, my Doctor of Science degree from the Massachusetts Institute of Technology (MIT) was heavily research- and theory-based, except for a half year of practical work at MIT's unique Chemical Engineering Practice School. I learned the art of chemical engineering process design while employed at the M. W. Kellogg Company during the war and then, coaxed by my longtime partner, Harry A. Rehnberg, launched a venture whose evolution I could never have imagined.

Initially, we performed contract design and engineering services for our clients' technology and invested our resulting fees into our own research. Our first success was a process for the direct oxidation of ethylene to ethylene oxide (then to ethylene glycol, an antifreeze and ingredient of polyester fiber). This process was eventually licensed to over thirty plants worldwide, and it naturally followed that we would manufacture the proprietary catalyst for these plants, many of which were also built by us. Other original processes were also developed and licensed. However, from virtually the beginning of our company's history, our goal was to enter into chemical manufacturing using our own technology. The reason was simple: in the words of bank robber Willie Sutton, "That is where the money is!" For a company with only $700 of paid-in capital, this was a vaulting ambition indeed, but one we had all planned and hoped for throughout our business careers. It was clear that we would need to find a partner, as opposed to just selling out our service-oriented business in order to retire. It is this ambition and its realization that I propose to describe in the balance of this essay.

CORPORATE PARTNERING

We learned early on to value more than just new technologies, discovering the importance of new institutional arrangements— social innovations, as it were—that would permit greater flexibility in the formulation of advanced and highly competitive technology. We had to invent many over the years, with corporate "partnering" our most significant achievement in social innovation. My experiences in forming the Oxirane group belong to this category.

In my view, corporate "partnering" fundamentally applies to the "cohabitation" of large and small organizations. If two large companies are organizing some form of collaborative organization, it would be commonly referred to as a joint venture; since both partners are perfectly capable, in most instances, of looking after themselves, there is nothing really new to say about joint ventures. But in the combination of a small and a large company, it is much too easy for the larger one to smother the smaller one. Our small company, however, was able to avoid such smothering, and both partners continued successfully together for many years.

Let me emphasize that I am not dealing with "intrapreneurship." This term signifies an effort on the part of a large company to create within itself a culture that is more like that of the entrepreneurial world and to keep out the suffocating culture of the large bureaucratic organization as long as possible. Some companies have managed to maintain "skunk works" for lengthy periods quite successfully. But there are many problems inherent in this organizational pattern, since I believe that the culture of a true entrepreneurial company is very difficult to sustain inside a large organization, no matter how good the intentions are on all sides.

One of the fundamental prerequisites for an entrepreneurial environment is the chance for great financial rewards; this creates an enormous amount of jealousy in a structured organization. At the same time, entrepreneurs recognize that they are taking a risk with their money and their future; personnel in a large company, even those assigned to "skunk works," rarely assume that the price of failure may be the permanent loss of their jobs. Decision-making in an entrepreneurial company is often done instantaneously, and a "committee" culture is seldom or never found. Work habits may be informal and the types of people employed are often quite different from those found in a typical large company. In general, smaller, owner-managed firms are better risk takers in many areas of activity.

Once, in a philosophical discussion with a representative from my corporate partner, the Atlantic Richfield Company (ARCO), we examined why our relatively small research organization consistently seemed to produce extraordinary results when many other similar organizations in large companies did not, despite the obvious advantages that they were offered. We attributed it in part to the fact that we had employees who lived close to the large metropolitan center of New York and who were intimately involved with global problems through the participation of the entire organization in world affairs. This gave them an extremely cosmopolitan perspective.

In private life, these people enjoyed many of the features that only a large city can offer. In fact, the kind of person who is attracted to New York is very often the kind who would not fit into the culture of a large company. For New York one can substitute few other major cities in the United States. Nevertheless, research organizations are often located on predominantly semi-rural or rural campuses where it is supposed that people can think more effectively. I have often felt that this is not always so and that the research environment for an industrial organization requires a feeling of intense energy—even pressure—and the knowledge that there are important problems that must be solved every day.

Finally, it seemed to us that a person who would be employed by a relatively freewheeling, smaller company would be more likely to be individualistic and hard to fit into a conventional organization. Many of our employees, in fact, had left large companies precisely because they wanted the relative freedom and excitement that they hoped to find with us. And we had a very good record of keeping most of our able people. These factors may have aided us in assembling a unique team of brilliant technologists, in a concentration seldom if ever found in larger companies. Many smaller companies today also have superb teams of specialists, and it is their concentration in the smaller company that can make fruitful partnering so rewarding for both sizes.

Although a successful partnership with a small entrepreneurial company brings many advantages to a large company, one may ask why a small company would want to be associated with a large company. Is it not afraid of being taken over, having its best ideas siphoned out and its best people enticed away? Of course, there are these risks, but written agreements can minimize such dangers. The large company offers capital, markets, greater resources and manpower, wider varieties of technological skills, and many other assets that would be virtually impossible for the small company to duplicate. The new technology that a smaller entrepreneurial company develops must be put to work as rapidly as possible in order to prevent obsolescence, achieve a position in the market, and pay for further R&D.

When first starting out, young, small companies like ours were usually forced to license to bigger companies in order to accomplish these ends, as some biotechnology companies do today. If this cannot be accomplished, smaller companies may sell part or all of themselves to larger companies (some may be foreign) in order to achieve a positive cash flow. However, this can truly become a Faustian bargain, both for the company and the country. In most cases, I have discovered that because technology is capital, it ought to be invested to receive the optimum return on the R&D that brought it into being, which means a share in the ownership. It was this conviction, formed as we saw our licensees increasingly willing to pay only for our successes and not our failures, that led us to search for ways to manufacture the chemical products that our novel processes had made so attractive to investors. As in all symbiotic relationships, if each partner has something the other needs, they can live together indefinitely.

PARTNERSHIP GLEAMS

In the early 1950s, when initial plant sizes were much smaller than they are today and capital requirements were much less, a partnership with the right large company seemed like the best avenue to pursue because a large source of external funding was required in a capital-intensive industry. New technology must be swiftly exploited commercially on as international a scale as possible in order to maintain or increase market share and gain the constant improvements that result from learning how to use the technology more efficiently than the competition.

Our first opportunity arrived in 1955 when we invented the Mid-Century process for the bromine-assisted oxidation of paraxylene to terephthalic acid, the main ingredient of polyester fiber. After some discussion, we entered into serious negotiations with Standard Oil Company of Indiana (now Amoco). When I proposed a corporate partnership, their vice president for research looked at me and said, "Ralph, that would be like the mating of an elephant and a mouse." Clearly it was not yet time for cohabitation. We sold our technology to Amoco and mourned our first lost opportunity at corporate partnership.

From another attempt at cohabitation—which involved buying into an existing joint venture with two other companies—we learned a great deal about the problems of organizing a corporate partnership. In particular, it seemed that no provisions had been made for the future expansion of the venture or its dissolution, thereby necessitating further negotiations each time any kind of capital investment was called for or the partners' interests no longer coincided sufficiently. Although the plant in question was located at the site of one of the partners' existing plants, many of the products were sold exclusively by another partner who consequently controlled the market; pricing the intermediates could not be done at arm's length because of varying market conditions.

PARTNERSHIP SUCCESS

There were many other problems, but each experience was helpful in preparing us for the moment when our third attempt at corporate partnership succeeded, this time with ARCO. One learns from one's failures as well as from the successes. ARCO was not the first possible partner but emerged as the best one after various explorations.

The technology involved this time was our discovery of a process for the production of propylene oxide, the principal ingredient of urethane foams and other polyurethane polymers. Why was ARCO interested? They had an entrepreneurial management in both the chairman of the board, Robert O. Anderson, and in the head of their newly formed chemical company, Robert D. Bent. They had ambitions to enlarge their chemical company rapidly but did not yet have all the necessary people or technical resources. Although they were doing research in an area similar to our own, we convinced them that we were likely to dominate the patent position, and this proved to be the case. We knew each other from previous contacts, and the reputations of both companies, I am proud to say, were ones of integrity and skill (these are fundamental prerequisites for making a corporate partnership work). Furthermore, it was a unique moment because the oil industry had a keen appetite for expansion in petrochemicals and a good raw material position to support it.

Our motivations were obvious. ARCO had all of the advantages large companies enjoy, and yet their new chemical division was sufficiently small and unstructured that decisions were reached much more rapidly than would be customary in a major oil company. It took us only six months to negotiate the agreement that was signed on June 30, 1966. This agreement governed most of our relationships for a long time thereafter, even though the top management of their chemicals group changed several times.

Some of the basic principles involved in our partnership with ARCO should be mentioned.

Management

We established a separate 50/50 corporation, the Oxirane Corporation, in Princeton, New Jersey, midway between the Philadelphia and New York offices of the parents. Charged with managing the business and technology of the parents in the defined fields, this corporation had an equal number of directors from each partner and functioned as a non-manufacturing entity primarily concerned with R&D, engineering design, and general management (including the provision of general advice to the manufacturing partnerships and overall marketing strategy).

A separate manufacturing partnership, designed primarily to build the first plant, was set up in Bayport, Texas. I use the word "partnership" here in its legal sense; the first and succeeding plants were indeed structured as partnerships directly by the two parent corporations through intermediate holding corporations, who were the actual partners on a 50/50 ownership basis. In this way, each partner was able to deal with the finances of the venture according to its own methods of accounting, and avoid the double taxation that usually takes place if a corporate tax-filing form is used. Joint decisions had to be made regarding depreciation, investment tax credit, and so forth, but each partner handled its respective share of tax by itself. Each partnership had a general manager and a chief technical officer.

Financing

Obviously, the financial agreement was the most important of all the arrangements. From the beginning ARCO recognized that we could not possibly carry an equal load in the financing of the enterprise, but they insisted that we own a 50 percent interest. In their view, this was particularly important because an equal arrangement of this kind would give us a real voice as a complementary equal in the affairs of the business and a strong incentive to exert maximum efforts on behalf of Oxirane, in addition to contributing technology rights. Consequently, it was they who proposed the basis of the financial arrangements that were incorporated into the original agreement. The basic features of this arrangement were as follows: *1)* The objective was to permit Halcon to obtain a half interest in Oxirane, while allowing ARCO to rapidly recover its investment and earn a satisfactory rate of return. *2)* ARCO provided 90 percent of the project's initial capital requirements; Halcon contributed 10 percent. *3)* ARCO's contribution was divided into two parts: 50 percent for its share of the venture and 40 percent as a nonrecourse loan to Halcon, to be used solely to fund its interest in Oxirane. *4)* ARCO was satisfied that the technical risk was acceptable. Had the venture failed, ARCO would have borne most of the loss. Despite the 90/10 ratio, however, any loss would have had a much greater effect on Halcon, by far the smaller company. The initial 10 percent funding was *very* earnest money for us, and that was what ARCO wanted us to express as a measure of our confidence in the technology and the project. *5)* The 40 percent of the total project cost (grass-roots capital cost) lent to Halcon was divided into eight notes. These notes could not be retired until at least three years had passed from when the venture first started and, even then, only one note could be prepaid every six months. The number of notes still outstanding governed the distribution of cash flow between the partners. *6)* The financing agreement was structured to maximize distribution; project cash flow would be allocated quarterly. Fifty percent of the cash went to each partner but if eight notes were outstanding, 40 out of Halcon's 50 percent was dedicated to servicing the notes. If seven notes were outstanding, the dedicated cash flow dropped to 35 percent, and so on at 5 percent per note. *7)* The notes had a nine-year amortization and a 10 percent rate of interest. If the 5 percent dedicated cash flow per note exceeded the sum of scheduled principal and interest, a portion of the excess was allocated to additional retirement of principal. Consequently, with very successful ventures, as the Oxirane plants proved to be, the principal was paid back substantially ahead of schedule, reducing the prepayment amounts. *8)* ARCO's rate of return was less than the project's return, although not by much. For a project having a pretax return of, say, 25 percent, ARCO's return significantly exceeded 20 percent because of the fixed 90/10 distribution in the first three years (in discounted cash flow analysis it is the early years that are the most important). As a private company, we were relatively unconcerned with "up-front" payout; we were interested in building long-term cash flows.

Some of the secondary issues should also be mentioned. If additional cash was required for plant modifications or working capital while there were still outstanding notes, Halcon could borrow an additional amount from ARCO basically by increasing the notes. If there were no outstanding notes, Halcon had to provide its full share; if all notes were outstanding, it could borrow 80 percent of its share.

As new units were added, it became necessary to layer the financings and to create methods for allowing Oxirane's operating management to make the correct decisions without worrying about the mechanics of the funding. For example, the second plant was much larger and more efficient than the first and was started up while most of the notes on the first were still outstanding. This was handled by creating a full-capacity financial model for the first plant and allocating cash flow as if the first plant ran at full capacity regardless of the plant's actual operations.

Personnel

Great attention was paid to populating the joint enterprises with the most qualified people from each partner. Benefits and seniority were set by the ventures, especially for third-party hires, but the key personnel from each parent were put on leaves of absence so that they could return safely to the parent companies at the appropriate time. Intermixing employees from two such divergent cultures made for an exciting and lively organization.

How the Partnership Fared

On the whole, the partnership fared very well. Our first plant started up on January 1, 1969; before it began, we were already building a larger second unit in Bayport and a substantially equivalent unit in Rotterdam in the Netherlands, which was financed in part by offshore bank financing combined with interpartner financing, a result of American currency restrictions that are now long forgotten. Simultaneously, we began negotiations in Spain for a plant to produce propylene oxide utilizing a variant of our technology, with the Spanish government as our third partner. This three-way partnership succeeded quite well because of the care with which the agreements were written.

After the second Bayport plant and the Rotterdam plant had started up, a third Bayport plant was commissioned, and a new agreement was entered into with two prominent Japanese companies. This agreement differed from the others because the Japanese agreed to arrange for all the financing and to give ARCO and our company 50 percent of the equity in return for the exclusive rights to the technology in Japan. This was a landmark decision by the Japanese government, and it has become, although somewhat modified today, the largest petrochemical joint venture in Japan. A substantial degree of the initiative in forming overseas ventures came from Halcon, which had twenty years of diversified international experience, while ARCO was then mainly a domestic oil company.

Neither partner had ever made propylene oxide before, which was a market served by at least twelve companies in the United States as well as numerous companies abroad. Dow Chemical Company was the leading producer. All of the existing producers employed the chlorohydrin process, which had been invented around the time of World War I. The possession of modern technology made it possible for us to acquire a substantial amount of the market growth in the products we made, and because there was excellent growth in the derivatives, we were fortunately never really involved in any major price wars to gain market share. There was no conflict of interest among the partners since neither one made the principal products; although one product was used by ARCO, which developed a wholly new market for it, it was easy enough to establish a market price.

Because of our extensive experience in plant design and scale-up of new technology, we were given responsibility for all engineering designs, although ARCO had final approval. Many of our R&D and design people were immensely valuable in plant operations as well as in general management. We found that plants that used new technology needed to be saturated with such highly technical people, who understood the technology and the plant design. In this way, start-ups were created rapidly and the learning process progressed quickly. This not only improved productivity, product quality, and profitability, it also reduced the chances of competitors entering the field. In fact, our experience in this area contributed to accurate and rapid estimates for the cost of the capital involved in other new projects.

By 1980, we knew that we had a tiger by the tail. The volume of business generated by our corporate partnerships on a worldwide basis had reached $1 billion per year. This was exceptional for a new chemical company that had made its first pound eleven years earlier. Our success was based entirely on original technology. But the time had come to give serious thought to whether we could stay the course indefinitely. For one thing, we now had to contend with inflation and high interest rates. The financing arrangement of our initial agreement had worked very well for the three Bayport plants and the Rotterdam plant and, as mentioned above, had presented no problem in Japan or, for that matter, in Spain. Our newest plant, however, which started up in 1978 in

Channelview, Texas, had cost many hundreds of millions of dollars. This plant was not financed under the original agreement for a variety of reasons, including the fact that ARCO was by now deeply involved in Alaska and the relative abundance of capital that we had enjoyed in the past was gone. We had to borrow money, secured only by the new plants and not by the parents of the new company. In those uncertain times we were at the mercy of a floating rate of interest, which soared upward. Obviously, Halcon counted on the cash flow from its previous ventures to sustain the investment in such future ventures.

THE END OF AMERICA'S DOMINANCE

For several years during the late 1970s, as inflation neared the 13 percent mark, interest levels remained subdued (meaning, of course, a negative real rate of return to the banks). In October 1979, President Carter was compelled to slow the economy and appointed Paul Volcker as chairman of the Federal Reserve Board. Soon thereafter, instead of targeting interest rates, as the administration and the previous Board had done, Volcker limited the growth in the money supply, which had been increasing rapidly and fueling inflation. In effect, the government had been inflating the currency by printing money. The prime rate shot up to 21 percent. Suddenly, all our cash flow was going to the banks. Technological strategy was no longer my principal concern—it was sheer survival. Could we meet the next interest payment?

PARTNERSHIP ENDGAME

At the same time, we had been involved with ARCO in renegotiating some of the terms of our original financing arrangement. I remember one of their financial executives saying, unimpressed by Oxirane's remarkable previous growth, "I will teach you the value of money" (versus technology, which had been the primary basis for our 50 percent share of the partnership).

It became clear that we could no longer sustain our position as an equal partner in Oxirane; ARCO's executive had been right. In such unfavorable economic conditions, money (finance) was decisive, technology was less valuable. Of course, some of the problems were ours. One of the plants at Channelview used a new process to produce ethylene glycol. While it eventually proved operable, it had been designed for an era of low energy costs, and those days were gone. We closed the plant rather than continue to invest in it in order to improve its performance.

Obviously, under these conditions, raising more capital from third parties was not feasible, nor could it have been negotiated in the short time before the next interest payments had to be met. We came to the conclusion, therefore, that it would be prudent to sell our interests. In June 1980, ARCO bought us out. It is worth noting that some of Halcon's employees elected to stay with Oxirane and progressed there; of course, they were free to do so. Most, however, returned to Halcon's more entrepreneurial environment.

Needless to say, we received a fair cash settlement for our many years of labor, and many of our employees benefited personally. The return on our research was spectacular. But ARCO ended up with a chemical business that is still a major part of the ARCO Chemical Company; I know that they have been expanding and further improving the status of this technology. In 1986, they announced the opening of a new plant in the south of France as well as enlargements at the Rotterdam plant, and new plants have just been announced for Europe and the United States. They have also increased the output at the Channelview plant. They are now the world's top producer of these products, and many of their competitors have discontinued their operations. New technology pays off if one is patient, entrepreneurial, and financially stable.

If ARCO and Halcon had not joined together, I doubt that today this business would exist and be as successful as it is. And I am not sure whether ARCO would have been able to accomplish the same result on its own. It is reasonable to ask, did ARCO treat us too generously in order to acquire our technology, our best people, and our entrepreneurial skills? Although some of their staff may have thought so in private, the great and continuing success of this business and the absence of analogous stories elsewhere lead us to believe otherwise. Only by this kind of arrangement could ARCO Chemical have become the successful company that it is today.

FURTHER MACROECONOMIC SHOCKS

In 1981, the second Volcker-induced recession occurred. By the summer of 1982, it was impossible to continue our remaining engineering licensing business, and we sold the still-independent Halcon SD group to the Texas Eastern Corporation. In 1985 our last remaining ownership (10 percent) in a major Brazilian petrochemical company was sold, just ahead of the virtual collapse of that country's economic strategy and the rekindling of inflation.

THE RISE OF FINANCIAL DOMINANCE

Why did this happen to such a technological triumph? Because I only dimly understood the causes of the traumatic events of the 1970s, I decided to join the economics community at Stanford and Harvard and study in particular the interaction between technology and economics. In the remainder of this essay I will briefly describe some of these interactions.

Managing for the long term requires CEOs and technologists to have a firm understanding of the changing business and technical environment with which both CEOs and technologists must contend. The events of the 1970s and 1980s have bred a more stringent financial climate that, understandably, has propelled many financially-trained executives to the top of their companies. Many of these executives know comparatively little about technology and have less time to keep up with the technological developments of their more technology-oriented competitors.

To the company's technologists the financial orientation seems to lead to a short-term mentality, but even a superficial knowledge of financial theory would suggest that, at high interest rates and cost of capital, the benefits from the gains of future earnings would be heavily discounted by financial analysts. This condition still does not relieve managements of planning strategically, and some, but not all, do.

One result of this increasing financial orientation among American companies has been that privately owned, high-risk start-up companies have had few problems finding venture capital (which was not the case when we first started) and planning for a longer time horizon than pure financial analysis might justify. When companies are privately owned, they often grow rapidly and have a strong technological orientation. The problems begin when capital is needed quickly, and the public markets must be tapped by initial public offerings. The entrepreneurs and venture capitalists who start these companies earn a handsome profit—the real reason why venture capital firms seem so much more patient than the banks or stock markets. The risks inherent in the enterprise are now spread over many investors. The psychology and motivation of management soon changes and becomes more like that of companies that have long been publicly held. Of course, these daunting circumstances can and have been overcome by superior technology, superb management, and farsighted entrepreneurs. Examples like Intel and Microsoft come to mind, but because they are so familiar does not make them any less unusual.

During the 1980s we saw some reverse buy-outs, where public companies, combined with leveraged buy-out specialist firms, went private in order to improve their longer range position, only to go public again when market conditions permitted. Private ownership fosters a longer term technological strategy, but the need for capital to grow or to cash in forces access to public markets in most cases.

When this occurs, we find that increasingly the stocks of large public companies are owned by pension funds and other similar funds. For instance, Alcoa, a Dow Jones index company (on whose Board I served), is almost 80 percent owned by such institutions. This leads to a growing separation of ownership and management. Company executives are fiduciaries for their stockholders, who are increasingly pension funds that are themselves managed by fiduciaries. Fiduciaries on fiduciaries—is it any wonder that horizons are short? How else can the performance of professional managements be gauged by their (temporary) owners, the pension funds, than by quarterly financial returns? Business is now so complex that these stockholders cannot readily understand what is happening. Unfortunately, some managements also encourage this behavior by judging their own pension managers strictly on their regular financial returns. Furthermore, some managements, perhaps unwittingly, contribute to the separation problem by rewarding themselves independently of their company's performance. Nevertheless, well-managed companies do find support for their stock, despite these obstacles.

NEW CLIMATE FOR FIRMS

Of course, many managers who grew up during the halcyon days of the 1950s and 1960s found themselves unprepared for the problems of the turbulent 1970s and 1980s. This was one of the reasons why technologists and managers seemed to view each other from different perspectives. Although the situation differed widely among firms, most had to contend with completely changed external conditions. For example, time horizons had greatly shortened (what is referred to as the quarterly financial returns syndrome), there was increased global competition, and capital markets were rapidly changing. The American financial and securities industries had invented a bewildering array of financial devices: program trading, options, markets, hedge funds, asset allocation and mutual funds, American depository receipts for foreign stocks, currency swaps, "junk" bonds, LBOs, and takeovers, to name just a few. The net result of all these systems is to increase the liquidity of the capital markets and internationalize them. At the same time, however, there has been a great reduction in the number of true long-term individual investors. Rather, the markets are now dominated by "bettors" who have no compunction about moving almost instantaneously in and out of financial instruments. Stock prices undergo sharp changes, sometimes overnight, often determined by computer programming. It is an unsettling environment for management and indeed for many employees who depend on stock options for added incentives.

In addition, there was a higher real cost of capital. During the 1980s, the financial and economic communities focused on a previously little-noticed phenomenon—the cost of capital, which is the return financial investors require to induce them to make investments. It appears that the average real risk-free cost of capital to American corporations (debt plus equity, which, when averaged, is called the cost of funds to the corporation) has been at least 5 percent higher than that of Japan's. When corporate managers calculate the pretax rate of return that a new project must earn before they are willing to undertake it—they call this desired return the "hurdle" rate—if a project's economics do not meet the threshold, it will not be undertaken. Hurdle rates vary across firms and types of projects and are set by managers based on many considerations, i.e., risk or strategic importance, not by the investors in the capital markets. The "hurdle" rate, therefore, differs from the cost of capital, which is the rate of return before corporate taxes that the firm must earn on new investments to provide shareholders and bondholders with the returns they require. These returns depend in turn on the required returns in the markets for debt and equity. The cost of capital to the corporation is thus the pretax return that a company must earn to deliver the cost of funds (debt and equity) to investors. It depends, therefore, on the tax rules governing corporations as well as on the cost of its debt and equity.

Thus, whereas nominal "hurdle" rates for risk-bearing investments may be 15 percent in the United States, they may be only 5 to 8 percent in Japan. Part of this differential is also due to the lower risk premiums afforded by the Japanese financial institutions (main banks, *keiretsu*, and the like), although their situation has recently begun to change, and it is possible that much of the previous differential has started to diminish. Obviously, projects that must pass a 15 percent hurdle will have to have a much more rapid payout than those at 5 to 8 percent; the latter can justify many more projects that eventually can drive American businesses and technologies out of the market or even out of existence.

The major reason for this difference is the much higher Japanese savings and investment rate. In Japan, gross domestic savings averaged 28 percent of its gross national product in the late 1980s, in contrast with only 13 to 14 percent in the United States. By the same token, Japanese aggregate domestic investment averaged about 24 percent of its gross national product in the 1980s, compared to 16 percent in the United States. Therefore, despite its heavy investment rate, which recently surpassed that of the United States in absolute dollars (from a smaller economy), Japan still had surplus capital that it exported to the United States.

The unfavorable balance of payments for the United States was the mirror image of this capital importation—that is, the Japanese obtained their dollars to invest in the United States by selling us more than they imported from us. In fact, investment opportuni-

ties in the United States for them in the recent past were so great
that, in addition to this source, they borrowed dollars from international money markets. Much as we comment on how unfair
Japanese trade practices are (and some are), the root cause of our
deficit as well as our higher cost of capital can be found in the low
savings and investment rate in the United States. This same high
cost attracted Japanese investment here, leading to the buy-out of
American firms with important technologies. American managements must contend with these stark facts, which are largely
beyond their control. In the 1990s, the Japanese invested less as
the economy dipped into recession, but their savings remained
high; they exported more capital while buying fewer imports,
which has led to our much deplored rising trade deficit with Japan.

Of course, some managements do much better than others when
faced with these obstacles. Increasing international competition
and the discipline of the financial markets are forcing more restructuring, cost cutting, and attention to the dynamic compara-
tive advantage that strategic management of technology can afford. Managements trained in the easier time of past decades must
learn rapidly to adapt. The task is often daunting. Japanese managements, trying to catch up with the United States and forced to
compete earlier in international markets, work hard to improve
their performance and to cut their costs. Furthermore, enjoying
the advantages of abundant and cheap capital and better trained
technologists, they are often willing to take the greater long-term
risks that their economic environment permits. Despite the current
Japanese economic and financial distress, it would be naive to
discount the ability of Japanese firms to be fierce competitors.
Indeed, while Japan is in recession, its firms have been investing
throughout Asia, which according to most economists is the most
rapidly growing area.

Technology has become much more sophisticated and difficult
to comprehend, even by technically-trained CEOs who cannot
keep up-to-date technically in the face of the many other constraints on their time. This is all the more reason why trust between managements and their technologists should be nurtured
and extended.

Other factors need to be considered, e.g., increased legal liabil-
ity, changes in accounting standards, volatility of exchange rates,

frequently changing and complex tax laws, etc., but this is not the place to do so.

It is my belief that a stable, steady, low-inflation, high savings and investment policy regime by government, with much less variability than in the past, is the essential prerequisite for firms and their managements to grow, innovate, and compete in global markets. It is at the firm level that competitiveness is determined, and if many firms succeed in a country, then that country's standard of living will grow at a satisfactory rate, with an adequate distribution of its benefits. It has been my intention in narrating this tale to underline the need for technologists to learn economics and business management, as rapidly as possible, because the climate today is not as benign as the ideal one that I have cited above. In today's high-tech world, a capable venture-capital company can often complement the entrepreneur's technical skills and concentration with its business and economic "smarts," but there may ultimately be a large price attached, usually a loss of control or a distraction of management's attention by the often rigorous requirements of public ownership. The best firms, however, will thrive under these conditions.

Elkan Blout

Polaroid: Dreams to Reality

P OLAROID CORPORATION WAS ORGANIZED in 1937 with Edwin H. Land as President, Chief Executive Officer, and Chairman of the Board. Edwin ("Din") Land remained the founding and guiding spirit of Polaroid for four decades, and saw its sales increase during that period from $142,000 to over $1 billion, its stock listed on the New York Stock Exchange, and its inclusion as one of *Fortune*'s 500 companies. If Land were alive today, he would be the person to write this essay. I do so because of my strong feelings for him and for the company, feelings I developed as a full-time employee of Polaroid in its formative years, from 1943 to 1962. I joined Polaroid as a research chemist and served in various capacities, including Associate Director of Research, Vice President and General Manager of Research, and member of its Operating Policy Committee. In 1962, I left the company to accept a professorship at Harvard Medical School, but I continued to be associated with Polaroid as a consultant.

Polaroid's early years were Land's early years; the company exemplified the characteristics of the man—inventiveness, determination, hard work, ability to communicate. Land, a product of a characteristic middle-class family background, was unusual principally in his intense interest in invention and in its practical applications. Entering Harvard in 1925, he left a year later to pursue his ideas on light polarization. In New York he worked on developing a plastic sheet for polarizing light. Succeeding in this,

Elkan Blout is Treasurer of the American Academy of Arts and Sciences and Senior Advisor of Science for the Food and Drug Administration. He was Vice President and General Manager of Research for Polaroid prior to becoming Professor of Biological Chemistry at Harvard Medical School.

61

he formed Land-Wheelwright Laboratories in 1933 to continue his research into the applications that might follow from his invention; he was also anxious to begin manufacturing.

Land envisioned using his large-area plastic polarizers to reduce headlight glare, a use likely to have a very wide commercial application. A demonstration of the practicality of the plastic polarizers induced a group of Wall Street investors, led by Kuhn, Loeb and Co., to invest $350,000 in the newly formed Polaroid Corporation in 1937. In those days, that was a substantial sum of money, and it allowed the embryonic company to begin intensive manufacturing, development, and marketing. These efforts, concentrating on polarized headlights, continued through World War II until 1947. However, in 1940, Land concluded that Polaroid's primary effort ought to be directed towards helping to win World War II.

During the war years (1941–1945), Polaroid's efforts—calculated in terms of personnel and total product sales—increased enormously. Some thirty-six men and women were employed in 1937; at the end of 1945, the corporation's personnel numbered more than nine hundred. During these years, the company conceived, developed, and marketed many new inventions, including optical plastics for military range finders, vectographs (a system using polarized images to visualize objects three-dimensionally), and a new type of heat-homing "smart bomb." Scores of bright, young scientists and engineers were recruited, and company sales ballooned, reaching $16 million in 1945. Then came the end of the war and with it a large change in the company's fortunes.

My own first meeting with Edwin Land occurred on a steamy August afternoon in Cambridge in 1943. I had come to the Department of Chemistry at Harvard University as a National Research Council postdoctoral fellow in 1942 and, reaching the end of my fellowship, was considering what to do next. Though I had two offers of academic positions, I decided I wanted to do something "practical," and had accepted a position with a pharmaceutical company in Philadelphia. At my going-away party, one of the most brilliant chemists I had ever met, Robert Woodward, apologized for arriving late to the party. He explained that he had been with a young inventor who ran a Cambridge company called Polaroid. When he told Land that he was going to a farewell party for a Harvard chemist, Land asked why the chemist did not come to work for Polaroid. That offhand remark led me to visit Polaroid the next day.

At our meeting Land inquired about my interests. When I told him that they were in the field of light, matter, and their interactions, he became very animated. Why did I not come to Polaroid rather than go to Philadelphia? The Philadelphia offer, I explained, was a very good one.

Land was not impressed. "I will offer you 50 percent more than the salary that they are paying you."

This surprised me, and while I realized then what an unusual man Land was, I had certain doubts.

"I would like to work on the problems that I am interested in," I said.

"You can work part-time on the problems that interest you," Land replied, "so long as you work some of the time on Polaroid's problems." I liked that idea, but was puzzled by how it could be put into practice. I suggested that if an assistant was assigned to work with me on my problems, I could spend additional time on Polaroid's. Land agreed immediately, and the deal was closed on that hot summer afternoon. Returning to our small Cambridge apartment, my wife and I started to unpack.

So began my more than fifty-year association with Polaroid. From the first, I was fascinated by this man who knew what he wanted, made decisions quickly, and was aggressive enough to do something that was very unusual in the industrial world at the time—provide an assistant to a new, young chemist and give him the freedom to do exactly what he wanted.

From 1943 to 1945 I did research on a new type of heat-stable polarizer for headlights, and contributed also to the development of plastic optical devices. There is no question that Polaroid enjoyed great success during the war, attracting a very bright and imaginative group of young scientists and engineers who invented and manufactured new products that helped the war effort and that might, in time, contribute to the future of the company.

However, a sharp decline in government-sponsored research followed the end of the war. As 1945 drew to a close, it also became disappointingly clear that the automotive industry was unlikely to accept Polaroid's headlight system. While it eliminated headlight glare and would therefore certainly save lives, the carmakers showed no great interest in installing such a system in their cars. The intensive research, development, and public relations efforts appeared to have been wasted. Polaroid (and I) learned a useful lesson: It was not enough to invent and develop, it was not enough to be "right"; it was necessary to invent and develop something the customer wanted or felt he needed. Typically, Land saw things a little differently. I remember one conversation with him during the early 1950s when I encouraged him to agree to undertake market research on a new photographic product.

"Elkan," he said, "we give people products they do not even know they want, so why should we invest in market research?"

That, so far as Land was concerned, was that.

With the abrupt decline of government contracts and the failure of the automobile industry to adopt polarized headlights, Polaroid suddenly found itself with few marketable products. Fortunately, Land had an idea in 1944 that would prove to be not only the company's immediate salvation, but the making of its future. Photographing the scenery on a vacation to Santa Fe in late 1944, Land was asked by one of his daughters, "Why can't I see the picture right away?" Characteristically, Land thought a minute and then said, "Why not?"

He returned to Polaroid and, with a few trusted assistants, started to develop what we now know as Polaroid instant photography. Initially, this was a very small effort, concerned only with developing the image-transfer principles, the mechanics, and certain of the concepts that were to be the solution to the grand idea of instant photography. Though many of us knew about the project, the few who were intimately involved were those Land felt he could work with on specific tasks. From 1944 until the end of the war, this was a low-level effort; it became Polaroid's major effort when the war and government support for defense-related research came to an end.

By 1946, Land and his small group had produced reasonable silver-image (i.e., black and white) photographs, but there were many difficulties with them. The images were sometimes unstable, and reagents often remained stuck to them. Also it was hard to control the process so as to make it work satisfactorily every time. At this stage the chemists were called in, and we helped produce chemical solutions to these very practical problems. Now an even more important and difficult task loomed: to convert a laboratory curiosity into a saleable product. At this time, the number of Polaroid employees had dropped to about 250 (less than one-third of the company's wartime strength) and the sales in 1946 were practically nothing. These were perilous times, but Land's enthusiasm for realizing his dream of instant photography never flagged.

Fortunately, the company had attracted a group of extremely bright and dedicated young people, including William McCune, who was recruited in 1939 as a quality-control engineer and was now leading the important camera-development operation; Charles Mikulka, a brilliant patent lawyer determined to protect the many inventions relating to one-step instant photography that were coming from Land and the others; Otto Wolff, a great originator of mechanical ideas and solutions; and Howie Rogers, a college dropout nurtured by Land's philosophy of inventive intuition, who came up with the fundamental concept of combining dyes and silver developers in one molecule for instant color photography. They and others of their quality worked endless hours for days, weeks, and months, finally solving many of the practical problems associated with the production of black-and-white instant photographs.

Land was the indubitable leader and focus of this effort in instant photography. The first instant camera went on sale in November 1948, and from that moment 90 percent of the company's efforts were dedicated to developing the field. Though a system was being marketed, there were still many problems. We did not know how to make the film reproducible, and the camera production was far from satisfactory. The first sepia pictures were followed quickly by true black-and-white photographs; the original, somewhat ungainly, camera was succeeded by other less cumbersome cameras. When the company began a very aggressive marketing of Polaroid-Land instant photography, the volume of sales increased, and profits began to reappear. Morale in the company picked up, and by the end of 1950 it was time to start thinking about the next generation of instant photography: instant color photographs. I was given the go-ahead to hire chemists for the research and development of the new chemicals that would be required, and the seemingly impossible task of producing color photographs one minute after the picture was snapped became a matter of the highest priority.

All of us understood that instant color photography required more than drive, enthusiasm, and good intentions. To achieve early production of the very complex color film, we needed help from an organization highly experienced in conventional color photography. Land had developed technical associations with some Eastman Kodak people in Washington during the war, and it seemed natural for Polaroid to turn to Kodak. Because a number of the high-level research people at Kodak were intrigued by the idea of producing *instant* color photographs, Polaroid quickly reached an agreement with Kodak: Polaroid would do the fundamental chemical and process research, inventing and synthesizing molecules that were both dyes and photographic developers, and Kodak would criticize our work and contribute its broad color chemistry experience, providing emulsions suitable for the color process. In return for this help, Kodak would secure the right to produce the negative film for the first Polaroid color photographic process.

During the next six years, starting in 1952, scientists and engineers from the two companies met each month, alternating between Cambridge and Rochester. It was a splendid collaboration—all the participants enjoyed the intense effort and learned a great deal. Many patents were issued as a result of this collaboration; almost all of the primary ones were assigned to Polaroid, including forty or so in my own name.

Although the Polaroid/Kodak meetings were intense, there was a camaraderie and lightheartedness that I have never forgotten. I remember one particularly grim evening in a drab hotel in Rochester. McCune and Land were both tense and were grousing about being obliged to travel to Rochester every two months when Land suggested to McCune, "Let's try some bed-jumping." "Bed-jumping" turned out to be a contest to see who could jump the highest and the farthest over the hotel's beds. The two spent the next hour jumping over the beds in our various rooms. Although Land was adept and competitive, McCune was six years younger and in much better physical condition, and he won. This sort of silliness, which we all enjoyed, provided some relief to the hard-driving, competitive work of the young group that gathered at Polaroid in the 1950s.

In Massachusetts, a new film plant was built in Waltham in 1954 to accommodate the growing sales of instant black-and-white film. At that same time, we completed the research and development that led to the production of three individual photographic images that would somehow be blended into one for a final full-color print. When I saw those monocolor images, I knew that we would eventually succeed with an instant color film. To support my hunch, I borrowed money for the first time in my life and bought Polaroid stock. It was a decision that I have never regretted, since the stock increased in value one hundred twenty-eight-fold in the next fifteen years.

By 1957, Polaroid was selling great quantities of instant cameras and even more instant black-and-white film. We had become a manufacturing and marketing company as well as one involved in research and development. Sales that year were over $50 million. The stock, listed on the New York Stock Exchange, began trading in November 1957. Although most of the key Polaroid employees were now reasonably well-paid, only a very few had an equity interest in the company. A discussion with my brilliant brother-in-law, whose business was investment, convinced me that morale and motivation at Polaroid would be stronger if a greater number of the top people were partial owners of the company. I suggested to Land and the Board of Directors that Polaroid consider adopting a stock option plan for certain employees. While there was some resistance at first, as well as some talk about the dangers of buying Polaroid stock, in the end the Board of Directors agreed with the proposal. In 1958, a stock option plan for ten key employees, including myself, was adopted. As I remember it, we were each issued options to buy up to 1,900 shares. For many years the stock did nothing, but eventually those options were each worth over $1 million.

During the 1950s we introduced several new cameras and produced and marketed new kinds of film, including some that were super high speed. Land continued to be interested mostly in the company's research and development activities and was always accessible to his top associates in those areas. I remember once in the mid-1950s when I went to his office to tell him about an important chemical development. Appreciating the significance of what I had described, he leaned back in his chair, looked at me, and asked, "Elkan, do you know what one of my dreams is?" I did not. "I will tell you," he said. "I would like to have in this office a barrel filled with cash, and when somebody comes into my office and tells me about a significant corporate accomplishment, I would like to reach into the barrel and hand him or her a fistful of dollar bills."

I laughed and said, "That is a great Land idea. When are you going to start?"

That was the end of the conversation.

By 1960, Polaroid's net sales of photographic products were at the unimaginable sum of almost $100 million a year, and the company had over three thousand employees—an increase of more than tenfold from our low point in 1945–1946. Perhaps more importantly for the future of the company, it was producing small quantities of its first one-step instant color film. To the scientific parents of this brainchild, myself included, it was all very wonderful and beautiful. By 1963, Polaroid color film was being widely sold—it was an instant market success—and I severed my ties with the company as a full-time employee.

In 1960, after our instant color photography had been proven in the laboratory, I told Land that I was thinking of leaving the company in order to return to an academic position. He thought about it for a while and did not seem surprised. Late in 1961 I went to his office and told him that I had been offered a professorship at Harvard. He confessed, "I knew both Columbia and MIT were after you, but I did not think much about it. Then I heard Harvard made an offer, and I knew we would probably lose you." He was right. I accepted the offer of a professorship at Harvard Medical School and returned to a full-time academic research and teaching career. Nonetheless, Polaroid asked me to remain as a consultant, and I did so for the next thirty years, having regular contact with Polaroid management.

My departure from full-time status at Polaroid gave me the opportunity to think about the company, Din Land, and his relations with my colleagues at the company. Why did I decide to leave the company at a time when it was doing well scientifically, technologically, and financially? At the time, I justified my departure by saying that the major challenge—the development of instant color photography—was over and there were no real challenges left at Polaroid. This was probably true for the time being, but certainly not in the long run: the company was amassing huge reserves, of both people and dollars. It only remained to harness them and direct them in the proper way. From the vantage point of thirty years later, I now think that I left because I would never have the identity at Polaroid that I could in the academic world, where what you are recognized for depends directly on what you yourself do. In a corporation, what you are recognized for depends not only on your own accomplishments, but also on your colleagues and your boss. There was never any question in anybody's mind that Land was boss at Polaroid. He was boss not only in the corporate sense, but also in the research area, and I suppose that became clearer as time went by. He was not only Chairman and CEO, but also held the title of Director of Research. Although I was Associate Director of Research and Director of Chemical Research, Land never relinquished his title of Director of Research during his active period at Polaroid, which indicated where his true interest lay; his research decisions would always be governing, never mine. Also, it was clear to other people in management that Land never liked to have anyone in the company who was clearly second in command. As he himself put it, "There isn't any number two; there are a lot of number threes."

All of us who worked closely with Land knew that he was extremely competitive, and he showed this in many ways: in the commercial area, where he was fierce about patents and became involved in two major litigations to validate Polaroid's patents; and in the personal area, when he played poker. His competitive nature was revealed to me in a very personal way around 1950. I had a parking space next to his in the Polaroid lot, and we often had conversations there about the company and other matters. He had a beautiful Oldsmobile convertible, and one day I said to him, "I love your car. If you ever want to get rid of it, keep me in mind."

About nine months later he called me out of the blue and asked, "Elkan, do you still want to buy my car?"

"Yes," I assured him.

"Okay, will you give me $800 for it?"

"Sure," I said.

"Fine," he replied. "The car is yours. Go ahead and pick it up in Central Square; it has a broken axle."

Was that winning? I enjoyed the car for several years after I had the axle fixed, and I thought the price was fair. Even with his competitiveness, Land was always sensitive to his colleagues' feelings. The Oldsmobile incident was not malicious; he thought it was a good joke, and in time I came to agree with him.

With Polaroid's original color film, you peeled the positive off from the negative. Although the color was beautiful and stable, the process was not functionally elegant, and Land quickly realized that he had not quite realized his dream of the camera that gave you an instant color print easily. We had often talked about the possibility of making an integral film in which it would not be necessary to peel apart the positive and the negative, and work in this area began in earnest in the mid-1960s. Land organized a small group to start working with him on how to make an integral film, and this group, as usual, worked in secrecy. Land was always most comfortable working with small groups of colleagues who would perform the experiments he requested quickly and efficiently and show him the results before a new experiment was attempted. This process worked well in many areas but was not really appropriate for developing the hardware—i.e., the cameras—necessary for the commercialization of the films. Bill McCune soon became Land's right-hand man in the camera-development area, although he was not always privy to Land's thoughts and private experiments.

The manufacture, marketing, and selling of both film and cameras were by this time the major part of Polaroid's business, with Bill McCune providing the vision and infrastructure in product development and manufacturing. He was the one who conceptualized the need for Polaroid to manufacture its own films; these manufacturing capabilities gave Polaroid the critical independence from Kodak. In addition, McCune constantly emphasized the importance of product quality, which he saw as crucial to product acceptance in the marketplace.

One of the main objectives in selling cameras was to increase the use of film, since film was the product with very high profit margins. For many years, Polaroid was the only company making instant photographic film; it was the kind of monopoly of which every inventor dreams. Despite the initial thrill of instant pictures, however, people seemed to tire of it, and the amount of film used for each camera sold decreased markedly as time went on. Some of us thought this was due, in large part, to problems associated with the cameras; others thought it could be traced to the quality of the resulting pictures. In any case, Land was eager to develop an entirely new type of camera, one that would have integral film— a camera with which you could just snap a picture, develop it within a minute, and see a finished print without needing to remove the negative from the positive or coat the picture to increase its stability. This effort—developing the SX-70 camera and its film—occupied much of the research, development, and engineering staff from 1956 to 1972. Land's enthusiasm for the SX-70 camera was boundless, but many top-level executives did not agree with his sales projections for it.

The SX-70 had many original features, including a new type of range-finder and new electronic light-monitoring circuits. It was introduced in 1972, and Land subsequently made the cover of *Life* magazine with the accompanying caption, "A Genius and his magic camera." Several components of the camera were originally supplied by outside manufacturers, but by 1973 SX-70 cameras were being assembled and sold at the rate of five thousand a day, and the manufacturing facilities that would allow Polaroid to take over the production of the entire camera were near completion. Although it was assumed that Kodak would be satisfied with the portion of the instant color photography market that we ceded to them in giving them the right to supply us with color film, this was not to be. Polaroid set up its own color negative manufacturing facilities, with McCune in charge of that area. Fortunately, he had recognized this need and organized an initial development group back in 1958.

By 1974, there was a good deal of discussion and considerable internal disagreement about the direction in which the company was going, which came to a head in 1975. Just after Christmas of that year, Land resolved these disagreements by asking McCune to be President and Chief Operating Officer. It is significant that he retained the title of Chairman of the Board, Chief Executive Officer, and Director of Research. This was a period when Land had been receiving many honors from the academic and industrial worlds, but it was also a time of substantial internal pressure to increase Polaroid's sales of both cameras and film, its most precious resource. Additional strain was put on by the salespeople, who asked for lower-priced products in order to increase their sales. McCune was the guardian of quality, both in cameras and in film, but his voice was often subjugated by and sometimes lost to so-called corporate needs.

There were also major attempts at this time to develop and market products other than instant cameras and film, among them document copiers and an instant movie camera, projector, and film. The latter project, called Polavision, was one of Land's favorites. It had begun rather modestly in the 1950s and in the 1970s was approaching public disclosure and sales. Not everyone in management positions at Polaroid shared Land's enthusiasm for the Polavision project. The objections were many, but foremost among them was the fact that this project's technical premise was based on additive color, which is inherently much slower than the subtractive color process used in one-step print photography. In addition, home video cameras were becoming widely available.

McCune took over as president in 1976, but Land continued as CEO, and there was no question that he was still *the* innovative force within the company. He was very good at getting people excited about his ideas and organizing small development groups to carry them forward. He had a real knack for determining what was new and important, but he was not interested in the day-to-day operation of the company; bringing together the disparate forces in management to work towards common, predecided goals was certainly not his forte. It fell largely to McCune to respond to the sales department's entreaties for new and cheaper cameras to bolster their sales.

Nineteen seventy-seven was a year of two major events for Polaroid. The first was that Polaroid's sales exceeded the billion dollar mark for the first time; the second was Land's demonstration of Polavision. Polaroid marketed Polavision the following year, and the reception was less than overwhelming. Compounding the corporate woes was a pending suit against Eastman Kodak for patent infringement; Kodak had introduced an instant camera and film in 1976 that was too much like Polaroid's. Pursuing, litigating, and resolving the suit became a major preoccupation at

Polaroid for the next ten years. Polaroid eventually won the suit and was awarded over $900 million in damages, which was then the largest settlement of a patent case.

Financially, 1977 and 1978 were very successful years due to increased sales of both cameras and film. For optimum quality, Polaroid cameras required a significant amount of high technology to be built into them. The salespeople were arguing for less expensive cameras in order to expand the market, but how could this be done considering all the technology required for good instant pictures? The answer, unfortunately, was a group of instant cameras that was not very satisfactory.

And then there were the problems with Polavision. Not only were sales disappointing, but a lot of inventory was languishing in the warehouse. In addition to the inherent difficulties of the product, there was a fundamental problem that was beyond Polaroid's control: instant movies were now available by using video cameras. Why did Polaroid seemingly ignore the immense competition of the video camera when it decided to go ahead with Polavision? Regardless of the answer to this question, which only a few people know, it soon became apparent to the management of Polaroid that marketing Polavision was a costly mistake. In 1979, after considerable anguish, the Board of Directors decided to write down the huge inventory of Polavision products. This decision was a major rebuke to Land by his own board of directors. Although this was not the first time that one of Land's innovative ideas failed to find a commercial success (the headlight application of polarizers being another), this was the first time that the company's fortunes and future suffered markedly as a result of the decisions he made. In early 1980, Land decided to step down as CEO and McCune took over that position.

By that time, sales in the instant photography field had slowed and profits, affected by the write-down of Polavision products, had diminished even more spectacularly. Where were the products and sales that would return Polaroid to the magic company that it had once been? Many people felt that improving the quality and reliability of instant color film was essential to increased sales; others championed cheaper cameras. A third group argued that new products were needed to respond to the new environment that Polaroid was facing in the 1980s. McCune was not only a great development person, he also possessed much of the vision and foresight that Land had shown in the 1950s. He felt that Polaroid's field was instant photography, and if that franchise could be combined with electronic imaging, Polaroid's position in the photography industry would be both consolidated and markedly strengthened. Unfortunately, his vision was not generally supported within the company, and the project was later dropped when McCune retired as Chairman of the Board. In my opinion, this was a major mistake.

Though the company's sales continued to grow in the early 1980s, its camera sales decreased, and it became obvious that either better marketing or new products were needed. Both strategies were attempted, but neither yielded satisfactory results. In 1982, Edwin Land resigned as Director and Chairman of the Board, and decided to devote all of his time to his scientific institute, the Rowland Institute of Science; the following year, he sold all his remaining Polaroid stock in a public offering. It was not a happy time for Polaroid. After winning the Kodak patent case, McCune resigned as CEO and was succeeded by I. MacBooth, a production engineer, until mid-1995 when the first "outside" CEO, G. DiCamillo, was elected by the Board of Directors.

From the mid-1980s to the mid-1990s, Polaroid's sales (adjusted for inflation) were essentially flat, as were its earnings. The company was very different from the company I joined in 1943. What precipitated these changes, the effects of which were seen in the 1980s and 1990s? Can a company comprised of intelligent, inventive people regain its former glory by reinventing itself? Or is it true that large corporations cannot maintain the vigor, drive, inventiveness, and dedication found in smaller companies? Should Polaroid have split into several smaller companies in the 1970s and 1980s when it was apparent that things were not going as smoothly as in previous decades? Perhaps a large organization should take its cues on how to succeed from biological processes, i.e., procreation and/or fission; maybe during the next years we will learn more about these and other tactics.

I have described some, but not all, aspects of Polaroid during the last fifty years. Prominent in this essay, and rightfully so, have been the character, personality, wisdom, and energy of its founder, Edwin Land. What was Land like? Knowing him was a unique experience. He was a true visionary; he saw things differently from other people, which is what led him to the idea of instant photography. He was a brilliant, driven man who did not spare himself and who enjoyed working with equally driven people. He possessed great powers of concentration. I have now come to believe that complete concentration is absolutely necessary for achievement. It is fine to be brilliant, it is good to work hard, but you must have real concentration in order to achieve in life. More than anything else, that is what I learned from Din Land.

As a manager, Land enjoyed recruiting competent people to work for him. Although a confirmed egoist, he was often good with people and was usually sensitive to their needs. He particularly enjoyed working with people who had little formal training—people whom he could shape, or, as he said, in whom he could bring out the best. Jerome Wiesner, a former President of MIT and a good friend and admirer of Land's, once said, "Edwin Land was a man who never had an ordinary reaction to anything." What more fitting tribute could there be to Polaroid's founder?

Gordon E. Moore

Intel—Memories and the Microprocessor

S INCE 1992, INTEL CORPORATION HAS BEEN the largest semiconductor manufacturer in the world when ranked on the basis of sales, achieving this position in less than twenty-five years. The company was formed in 1968 to make complex integrated circuits. Its annual revenue in 1995 was over $16 billion; in 1995 the world market for semiconductors was estimated to be about $145 billion.

Semiconductor technology is the basic technology used in modern electronics. It is used to produce computer chips in personal computers and in a wide variety of other devices such as televisions, cellular phones, light switches, kitchen appliances, and automobile engine controllers, to name just a few. These chips consist of intricate microstructures, which are etched layer-by-layer in complex patterns on the surface of wafers. These wafers are round, thin slices cut from hyper-pure, single-crystal silicon that is derived from common sand. It is truly a broadly applicable and ubiquitous technology.

The semiconductor industry began in 1947 with the invention of the transistor at the Bell Telephone Laboratory. A transistor is the building block of digital logic and memory circuits. It works much like an electric light switch: current is allowed to pass when the switch is in the open position; the flow of current is stopped when the switch is closed. Transistors from early on had a revolutionary impact on society because they made products such as low-power portable radios possible, bringing worldwide communications to previously isolated individuals.

Gordon E. Moore is a Founder and Chairman of Intel Corporation.

The first commercially available silicon transistors were introduced by Texas Instruments in the mid-1950s. The integrated circuit, the next revolutionary step in the industry, became commercially available in the early 1960s. Here a complete electronic circuit, including interconnecting wiring, is built in and upon a single silicon chip. By today's standards, these early circuits were relatively simple; they utilized at most a few transistors, resistors, and capacitors. However, they demonstrated the advantages that could be obtained by interconnecting individual circuit elements, using processes similar to those used to make transistors. The complete function could thus be made under the carefully controlled manufacturing conditions required by the semiconductor industry.

A unique aspect of the semiconductor industry is that prices for products tend to decrease over time. A 20 to 30 percent price decrease per year is about average, although this average consists of periods of time when prices fall very rapidly and when they might even increase if supplies are tight. Not only does the price fall for a given integrated circuit, but as the complexity of the chip increases, the price per electronics function decreases from product generation to generation as more and more functions are integrated into a single structure. Today a complete circuit containing several million transistors costs less to the user than did a single transistor thirty-five years ago.

This phenomenal increase in industry productivity is the driving force that has made semiconductor electronics the technology of choice for all control and computing applications. There are two reasons why the cost of this technology consistently drops: first, the broad applicability of semiconductor devices leads to a phenomenally elastic market, so that decreases in cost are more than balanced by increases in the total unit demand. The total market continues to grow in dollars despite the fact that the products are falling rapidly in price. Second, because of the unique nature of the technology, by making things smaller the speed of the circuits increases, power consumption drops, system reliability increases significantly, and, most importantly, the cost of the electronic system drops. An analogy to real estate is helpful. Our industry sells an area on the silicon wafer for about a billion dollars an acre; this has remained roughly constant since the advent of the integrated circuit. By making things smaller, development density is increased.

More function can be built on a given area, causing the price of electronic functions to be cheaper and cheaper.

This continued rapid evolution over the course of many years causes severe industry dislocations. The semiconductor industry gradually incorporates the tasks previously done by its customers. In moving from the transistor to the integrated circuit, the semiconductor industry assumed the job of circuit design. With the construction of more complex logic blocks, logic design was assimilated. Increasingly, whole subsystems, such as today's microprocessors—the "brains" of PCs—are supplied as standard building blocks. Some software is even being incorporated onto the chip as functions, previously a part of computer programs, are etched into solid-state circuitry. For example, floating point arithmetic—important in a variety of scientific and engineering calculations—is now done directly by the hardware rather than by software subroutines.

Dr. Robert Noyce and I left Fairchild Camera and Instrument Corporation in 1968, shortly before we founded Intel. At that time, Fairchild was a major player in the integrated circuits industry. Fairchild Semiconductor Corporation, which was acquired by Fairchild Camera and Instrument Corporation in 1959, was formed in 1957 by a group of eight scientists and engineers. The group's goal was to develop and market a new type of silicon transistor that could be produced in batches rather than individually. The founders came from Shockley Semiconductor Laboratory, a research organization formed by Dr. William Shockley, one of the three inventors of the transistor at Bell Labs. Dr. Shockley initially focused on developing a silicon transistor, but eventually shifted direction. The group founding Fairchild Semiconductor returned to the goal of a commercially viable transistor.

Fairchild's first products were mesa transistors, a significant improvement over silicon transistors available at the time. (The term mesa comes from the flat-topped microstructures that resemble on a microscopic scale the mountains of the Southwest United States.) More importantly, shortly after the first products were introduced, Fairchild engineers developed the planar transistor, a structure wherein the sensitive portions of the transistor were protected by an adherent glassy layer of silicon dioxide on a flat, or planar, surface. This new approach became the basis for making practical integrated circuits, and Fairchild was the first to produce them commercially. Integrated circuit technology advanced rapidly from the first simple circuits produced in the early 1960s, and the complexity of integrated circuits grew accordingly.

By the late 1960s, the technology had developed to the point where it was possible to make circuits containing hundreds if not thousands of components on a single chip. But the problem lay in defining functions of that complexity that would be useful when manufactured in high volume. The semiconductor batch technology is especially suited to making large quantities of identical structures. The cost to design and to build the complex mask set necessary for an integrated circuit is high, making the economics that allow cost reductions of electronics dependent on the ability to amortize this cost over a large number of units produced.

The industry was caught on the horns of a dilemma. We were capable of manufacturing circuit functions more complex than those that could be defined. The industry evolved to the point where the cost of producing the simple circuits for which there were markets, that is, the general purpose building blocks used by the industry, was essentially the cost of the package that contained the chip rather than just the chip itself. It could far exceed the cost of the tiny silicon chip. The semiconductor industry was fast becoming a packaging and micro-assembly industry, rather than one where product costs were controlled by the high technology silicon processing. Having large factories in low labor cost areas was becoming a more important competitive advantage than the silicon technology itself.

During this period, some top management problems developed at Fairchild Camera and Instrument Corporation, the parent company of Fairchild Semiconductor. Two chief executive officers left within a six month period in 1967, and the board of directors established a three-man committee to manage the company while they were searching for a new CEO outside the company. Bob Noyce, a vice president and the most logical internal candidate for the position, was clearly being passed over. Given the circumstances, Bob decided that he would leave Fairchild. At the time, I was director of research and development, running the laboratory at Fairchild. I had become increasingly frustrated with the difficulty involved in transferring new products and technology into the manufacturing organization of the company as the semiconductor division grew and became more successful. It seemed that it was much easier to get new technology picked up by groups that would spin off and start a new company than it was to transfer it to our sister operations. Technology transfer between groups is always difficult, but in the early days of Fairchild, when the technical competence clearly resided in the laboratory, it had been much easier. As the receiving organizations became increasingly competent, the difficulty of transfer grew. Partially as a result of this, we had a plethora of spin-offs. The areas of technology that we were working in offered a wide variety of product opportunities. This, combined with the development of the venture capital industry, led to a large number of new companies being formed out of Fairchild.

The fact that new ideas were spawning new companies rather than contributing to the growth of Fairchild was immensely frustrating. The rich vein of technology was clearly too rich for a company the size and breadth of Fairchild to fully exploit. Under these circumstances, it was reasonable for me to consider something new.

THE FORMATION OF INTEL

When I heard that Bob Noyce was leaving, I considered my own situation and decided that it would be best for me to make a move before the new management arrived rather than after. I told Bob that I, too, would leave and that we should try to find something to pursue together. When I told Dr. Andrew Grove, the assistant director of the laboratory, that I was leaving, he indicated that he too would like to join any operation that we were starting.

After leaving Fairchild, Bob Noyce informed Arthur Rock, a venture capitalist friend from the days of the formation of Fairchild Semiconductor Corporation, that we were planning to start a new company. Bob asked if Arthur would take on the task of raising the necessary capital. To get us going, Bob and I each invested some of the money that we had earned from the Fairchild start-up, but our $500,000 had to be supplemented by an additional $2.5 million to give us enough to really begin operation. Arthur undertook the task. He called several of his friends and on the basis of his recommendation got immediate commitments for the start-up financing. This was a period when venture capital was relatively available. Tax laws were favorable and the climate for success was good. In fact, we did not even write a business plan in order to get the financing committed. A one-page letter from Bob to Arthur that vaguely described the areas in which we were going to work was sufficient to get the initial commitments.

The goal of our company was to exploit the possibility of building very complex circuits. We saw in semiconductor memory a business opportunity that could change the leverage in the industry and return the critical capability to the careful processing of silicon. Memory is a function that is used broadly in all digital systems. At that time, it was performed principally by an array of magnetic cores—small doughnuts of ferrite material that could be magnetized in one direction or the other to store a 1 or 0. These ones and zeros were read or written by proper electrical pulsing in wires strung through the holes in the doughnuts. Memory systems using these magnetic cores of a million bits or so then cost a penny or two per bit. The closest thing to a semiconductor memory that existed at that time and was available commercially was a dual flip-flop circuit that sold for a few dollars. Thus, to be competitive in cost with magnetic memory, we had to see a cost reduction of over a hundredfold. We thought evolutionary advances in semiconductor technology offered the opportunity for this kind of cost reduction.

To get started, we found a facility where a small semiconductor company had once operated. "Silicon Valley" was full of small semiconductor companies, many of which could trace their origin to Fairchild. The building had many of the utilities that we needed and allowed us to avoid some of the start-up costs and delays that would otherwise have been required had we moved into a completely empty building. We hired a staff of young, highly capable people. At Fairchild we had learned that one of the advantages of a start-up is the opportunity to train managers. We hoped that we could recruit a staff that would grow in capability as fast as the new company grew.

There are basically two different types of transistors that are used to make integrated circuits: bipolar transistors and Metal Oxide Semiconductor (MOS) transistors. Bipolar transistors were the kind that Fairchild first made and were the ones used in early integrated circuits. In 1968, bipolar offered speed advantages over MOS at the cost of higher power consumption and lower packing density on the silicon, i.e., fewer transistors per acre. The MOS transistor was relatively new, but because of the high density with which it could be placed on the silicon surface, it offered the opportunity of lower cost.

At that time semiconductor memories were not available commercially, although IBM was using small memory chips in its big computers for temporary storage, what is called scratch-pad memories. These chips, however, were not generally available to other computer manufacturers. We decided to develop two new versions of semiconductor processing technology (one for each type of transistor). One was aimed at making a new kind of bipolar transistor that offered even higher performance potential, the so-called Schottky diode-clamped bipolar transistor, and the second was a new version of the MOS transistor technology called silicon-gate MOS, where the metal electrode used in earlier versions was replaced with a silicon film. This new MOS process offered the advantages of self-registration (meaning that one layer of the structure is automatically aligned with those previously applied to the wafer), so the devices could be smaller and perform at higher frequencies. It gave us greater flexibility in designing the interconnections among the various layers. The Schottky bipolar technology worked very well. Although we had anticipated some problems, the technology succeeded immediately. In fact, it worked so well that Intel's first product, a 64-bit memory, was produced in less than a year. Unfortunately, because the technology worked so easily, other established semiconductor companies had no difficulty duplicating it. The advantage that we had in the bipolar area did not last long, and eventually we dropped this approach. And so began what in retrospect I call our "Goldilocks" technology strategy.

We also pursued a third technology, a multichip assembly technology in which several silicon chips could be combined in the same package, thereby increasing density and, we hoped, lowering packaging costs. Today, we are still working on this approach at Intel. It has proven to be much more difficult than we anticipated, creating far more problems than it has solved. Had this been the only thrust of our start-up, we would have run out of money well before we got our first product to market.

Fortunately, the silicon gate MOS technology proved to be just right. Our first MOS product was introduced shortly after the first bipolar memory. When the established semiconductor companies tried to duplicate it, it took them longer than we had predicted.

Initially, we had hoped to have about five years to get established as a company before we had any direct competition from the industry giants. In fact, we had about seven years before they were able to successfully duplicate the MOS technology. That time allowed us to firmly plant our roots. The silicon gate MOS technology proved to have just the right degree of difficulty to allow us to start-up successfully and still establish a sufficient technological barrier to let us grow relatively unchallenged.

THE START-UP PERIOD

Start-ups are an exciting time. We divided the labor of establishing the technology and developing the first products among the senior people. The technology group pledged to develop the rudimentary processing before the end of the year so that we could proceed with product development. This pledge became a bet. To win the bet, the process people would have to get the equipment to operate reasonably well, make a stable transistor, and demonstrate a couple of other technical milestones by the end of 1968, only five months after start-up.

Moving into a building with no equipment on August 1 left relatively little time to purchase the process equipment, install it, get it running, and demonstrate at least a rudimentary process. The last of the three milestones was achieved on December 31, but only with the very dedicated effort of the entire staff. The camaraderie that develops in a start-up makes this kind of operation the rule rather than the exception.

The first product that we developed was a 64-bit static Random Access Memory (RAM) using the bipolar technology aimed at high speed memory systems, much like those that IBM was using in their computers. Sixty-four-bit was chosen as about the limit of where we expected to get good manufacturing yields. Any small defect can ruin an integrated circuit. The larger the circuit, the higher the probability of incorporating such a defect. We were pushing the limit. With six components per memory cell, this translated to about 450 transistors on the chip, far more complicated than the commercial products being produced at the time. For our first MOS device, we chose a fully decoded 256-bit memory chip containing well over a thousand transistors. It again represented the state of the art in complexity in

1968 (by comparison, the Pentium® Pro processor Intel introduced in 1995 contained 5.5 million transistors).

The first year was hectic since we did both product design and technology development while putting the framework of a sales and marketing organization in place. As we achieved major milestones in the development of these first products and the associated processes, we celebrated with impromptu parties. Typically someone would run down to the corner store and buy several bottles of champagne and enough ice to cool them in a wash tub. The entire staff would gather in our little coffee room to celebrate. By the time we outgrew this facility, the tiles in the ceiling were peppered with the imprints of all of these champagne corks.

There are certain advantages in being first with a new product concept. For example, when we designed the 256-bit MOS memory called the i1101, we designed it to operate at power supply voltages of +5 and +12v. When we built the product, however, we discovered that 12 volts was more than the device structure could handle; it operated better at +5 and +9v, so we changed the specifications. Since no systems had yet been built to use the memory, power supply voltages could easily be adjusted to what was required. Someone likened this to a rifleman who shoots at a blank wall, finds the bullet hole, and then paints the target around it. He always hits the bull's-eye. A second rifleman, however, finds the target already in place and therefore must hit the bull's-eye.

Our first products were introduced a year after our formation—the 64-bit bipolar product in August 1969 and the 1101 in September. Each was announced with a major advertisement in *Electronic News*. These products hit a relatively receptive market, but they still did not approach cost competitiveness with the established magnetic cores. Instead, as planned, they filled some market niches. The bipolar one filled the need for very fast, small memories, while the MOS product was competitively priced with magnetic core memories in small memory sizes.

There were, however, some problems with these memories compared to those currently in existence. Magnetic cores, for example, are able to retain information even when the power source is removed. The magnetization is permanent and can only be altered by an electric pulse especially designed to reverse the magnetization. With semiconductor memories like the i3101 and the i1101, "static" RAMs made with semiconductor devices, the information is stored in a flip-flop circuit. Once power is removed, the circuit rapidly returns to a neutral state and the stored information is lost. This volatile kind of storage has since proved to be very useful, but it was a source of concern for many designers of electronic systems in Intel's early days. It was certainly one of the reasons that the market for the first products was of modest size.

These products established Intel as more than a research and development operation; we became a manufacturer, capable of supplying the products in production quantities. Our detractors were inclined to pass off our products as laboratory curiosities, not as merchandise that could be supplied reliably. To try to erase this image of being only a research and development company, we made sure that we could satisfy all of the orders that we accepted.

The product that first challenged magnetic cores, that was extremely important in establishing Intel as a viable business, was a 1K (1024 bit) memory chip that we dubbed the i1103. This was the first dynamic random access memory or DRAM to be sold commercially. If the earlier static RAMs were volatile, this one presented even more of a system challenge. The idea was that information would be stored as a charge on a very small capacitor attached to a MOS transistor. However, if the memory was left alone, even with the power on, for a significant fraction of a second, the charge rapidly drained to the point where the information was lost. The word "dynamic" in DRAM refers to this aspect. In order to retain information, it is necessary that this type of memory have the information refreshed on a regular basis, approximately every one-thousandth of a second. This may sound like a chore, but since information can be addressed every microsecond or so, it means that only one one-thousandth of the time would be required to refresh any cell. The memory was refreshed in columns so that the array required some thirty-two cycles out of every one thousand or so, about a 3 percent performance hit, in order to maintain the information. This process gave at least a fourfold cost decrease over the static memory. When the price of 1000-bit chips could be reliably forecast in the ten dollar range, systems at about the same cost as those built by cores could be anticipated. By the time we introduced the i1103 we could make such a forecast.

While the trade-offs were different, the semiconductor memory offered other advantages over magnetic cores. It was easier to get short access times for high performance computing systems. Memory systems were relatively smaller, so more memory could be placed in a given volume. More importantly, the technology was clearly moving in a direction that would continue to cause the cost of semiconductor memory to decrease. Most major computer companies found the trade-offs attractive and undertook programs to use dynamic random access memory in their systems.

One problem, however, was that Intel was the only supplier of these products. We were still a small start-up company when the 1103 was introduced in 1970. It was conventional in the electronics industry for systems manufacturers to have multiple sources of any components because the suppliers occasionally lost the recipe and thus the ability to deliver. Also, the leverage on pricing that accompanied multiple sourcing was an attractive feature. For Intel to develop large volume markets for the 1103, it was necessary that there be an alternate source. The economy during this period was relatively soft. In 1970 and 1971, demand slowed significantly, and we began to worry about Intel's finances. Fortunately these two problems were handled with a single solution. Microelectronics International Limited (MIL), a subsidiary of Bell Northern in Canada, had been established to enter the semiconductor business. MIL's management was very interested in getting access to Intel's silicon gate MOS technology. They were intrigued with the dynamic memory and its potential for broad applicability. During this period of relatively slow business, we negotiated to transfer our technology to MIL to establish a second source for the 1103. In return, we received expenses plus a relatively significant fee, enabling us to maintain our operation through the industry's soft period. The contract included a strong incentive for Intel to transfer the technology rapidly and effectively to the MIL facility in Ottawa. In order to accomplish this, we had to move some of our people, accustomed to California sunshine, to Ottawa, where they discovered the harshness of the Canadian winters. MIL's first production line came up on the technology very well. In fact, for a period of time they were running higher manufacturing yields than we were at Intel, and they offered a viable alternative supply to semiconductor memory users.

MIL's line ran fine for some time, but then as the volume market grew, we at Intel decided to change our production from 2-inch silicon wafers to wafers 3 inches in diameter in order to lower the unit cost and increase capacity. Having been through previous wafer size conversions in the past, we took the challenge very seriously. We devoted a tremendous amount of energy to developing the new processes required for making the transition. However, while MIL had learned the 2-inch process well, they had no past conversion experience to guide them. It seems that we had transferred the 2-inch technology very effectively to MIL, but not the know-how to recover the process should something send it off course. When MIL tried to convert to 3-inch wafers, their yields collapsed, and they never really got back on track.

From Intel's point of view, MIL had been the perfect second source for the 1103. They did an excellent job confirming the availability of an alternative. By the time our customers finally needed the product in volume, we were ready to deliver as essentially a sole-source supplier.

Even then there were significant challenges to overcome to replace cores in most memory systems. Paramount, of course, was cost. The threshold for conversion was a memory systems cost of around a penny a bit. Thus, driving the cost of the 1024-bit 1103 well below ten dollars was crucial so that the complete memory systems cost could be under the critical value. The dynamic nature of the memory offered a challenge. Volatile memory was new for computer main memory. It took some real soul searching on the part of the system designers to choose a memory where the information was lost every time the power was shut down. And, of course, there were the usual technical problems. Some of the early memories tended to have disturb problems, that is, when information was written in one cell, the information in adjacent cells occasionally was destroyed. In addition, it was one of the most difficult to use integrated circuits ever produced. To build successful memory systems with the 1103 required considerable engineering expertise on the part of the user. This last problem may, in a perverse way, have been helpful in getting DRAMs established in the larger computer companies. Engineers with experience in core memory design initially saw semiconductor memory as a threat to their careers. Problems such as write-disturb gave them some assurance that their skills would still be required, because they had dealt with similar problems when they engineered successful core stacks.

Higher performance and the modularity of memory to much finer levels were advantages that all system designers could appreciate. Still, broad acceptance was slower than we desired. Intel decided that to accelerate acceptance we would establish a Memory Systems Division. To do this Intel would have to acquire the engineering and marketing talent necessary to design complete memory systems and to take these to market. While it was a major step for our still small company, it accelerated the broad adoption of semiconductor memory based on our early DRAMs.

There were two basic reasons behind Intel's formation of its Memory Systems Division: *1)* to expand the market by removing the need to engineer systems by those companies that were too small to have the internal capability and *2)* to address the largest single market for memory by selling complete memory systems that could be added on to IBM computers. To make memory systems we had to increase the range of our engineering skills far beyond our expertise in semiconductor chips. On the other hand, by acquiring these skills, we also gained insight into some of the problems that our customers encountered with our chips. This served us well in the design and development of future DRAMs.

The market for the 1103 grew rapidly, and the DRAM soon became Intel's most important product both from a financial point of view and from the impact that it was having on the industry. It was the first chip to challenge the established memory technology in its area of greatest strength. Also, at its peak, it was the single largest-volume semiconductor product when measured in revenue.

There were, however, other memories of importance in our early history. The static RAMs, the 1101 and the 3101, served relatively small niche markets. The evolution of these products continued, and by 1974 we were supplying a 1K-bit static RAM based on NMOS silicon gate technology (a major advance in performance for the MOS). Our products up to that time had been based on PMOS, a somewhat easier but lower performance process. NMOS allowed us to make a 5 volt only static RAM with high performance. The i2101 became the memory of choice for small systems where a static RAM was appropriate.

Another memory invented early in Intel's history was the Erasable Programmable Read Only Memory or EPROM. This product grew out of our efforts to build a non-volatile semiconductor storage device. We had observed a phenomenon at Fairchild that when an ordinary MOS transistor was biased in a certain way, charge seemed to transfer to the gate of the device even if it was not connected, turning on the transistor (a gate is a trigger that toggles a switch in a MOS transistor). This charge stayed in the device for an extended period. We had no idea how this happened. Dr. Dov Frohman, the engineer assigned to this project, developed a model for the physics involved in the phenomenon. He demonstrated that the charge could be removed by exposing the device to ultraviolet light, thus erasing any stored information. He invented a device that could be programmed electrically that would retain the information—for many years, as our subsequent tests proved—but which could be erased by exposure to ultraviolet light.

In February 1971, Dr. Frohman presented a paper on this device at the International Solid State Circuits Conference in Philadelphia. During the presentation, he demonstrated that the device could be programmed to different levels and that the ultraviolet erasure could be complete. He showed a display of the 2048 bits in Intel's first sample EPROM, all initially programmed. When the device was exposed to ultraviolet light, its storage cells gradually lost their charge until only those in the shape of Intel's logo remained. On further UV exposure, the remaining pattern started to fall out bit by bit until, finally, when the last bit in the field lost its charge and the display cleared, the audience burst spontaneously into applause. Dov received the award for the best paper at the conference.

In the first half of 1974, Intel's profit before tax exceeded 40 percent. Our business was based principally on this family of memory products: the i1103, the i2102, and a 2048-bit EPROM, the i1702.

THE MICROPROCESSOR

While developing our memory products we began to search for other opportunities to build complex Large Scale Integrated (LSI) circuits that could be used in high volume. An alternative to finding a general purpose function was to find a function that, because of the kind of system in which it was used, was built in very high volume. Electronic calculators were becoming popular and met our criteria for such a high volume application. We met with several calculator manufacturers to see if we could find one who wanted us to develop and supply the appropriate chips. Though the principal calculator companies had already developed partnerships with other semiconductor companies, we found a Japanese company, Busicom, that wanted to enter the market with a family of sophisticated business and scientific calculators and had not yet enlisted a chip supplier. In fact, they already had done the necessary logic design; they had thirteen complex circuits and all they needed was a semiconductor company to build the chips for them.

We soon realized that undertaking thirteen circuits of this complexity with no regular structure was far too much for a company the size of Intel. In 1969, when we were evaluating the Busicom opportunity, our engineering staff was already stretched to develop the few memory circuits upon which we were working. Marcian "Ted" Hoff, one of our young engineers, however, had considerable experience with general purpose computers. In fact, we had hired him out of his postdoctoral position at Stanford University because of his knowledge of computing systems. He saw that a general purpose computer architecture could be used to build all the Busicom calculators. He suggested that a processor, if designed to use transistors efficiently, could be built on a single chip of about the complexity of our memories. He envisioned a central processor chip, which when combined with a chip containing program memory storage in a fixed memory called Read Only Memory, or ROM, and a read/write memory, a RAM of the kind that we were developing, would make a complete computing system that could be used to make any of the Busicom calculators, changing only the stored program.

Ted not only saw the applicability of this approach to calculators but he realized immediately that it could perform a variety of control applications by utilizing a different stored program in the Read Only Memory. I remember him suggesting controllers for elevators and traffic lights as examples of potential applications.

We now had a way that we could work with Busicom, cutting the development effort from thirteen complex chips to one plus a couple of memories, and at the same time we could have a product of far more general applicability. First, however, we had to sell the idea to our customer. Upstart Intel asked them to throw away the design that their engineers had painstakingly created and adopt our revolutionary approach. We also asked them to pay part of the development costs. We were genuinely surprised when Busicom's chief engineer agreed without hesitation. After about an eighteen-month development effort, the first microprocessors were shipped for revenue in February 1971. This was a landmark in the history of Intel and, for that matter, in the evolution of electronics.

The era of broad use of the microprocessor did not begin immediately, but after repurchasing the rights to the 4004 processor and associated memory circuits from Busicom, Intel announced its general availability with an aggressive advertisement in *Electronic News* that proclaimed "A New Era in Integrated Electronics." Our product was the beginning of something that would have a significant and lasting impact.

MEMORIES AT INTEL

Memories remained our largest volume and most profitable product line for many years, despite the fact that other companies entered the market as the DRAM applications grew. Intel introduced a 4K (4096 bit) memory as soon as the state of our silicon processing technology allowed us to increase complexity and still maintain acceptable manufacturing yields. Our first 4K was an extrapolation of the 1103. One of our competitors, another start-up by the name of Mostek, took a slightly different approach that had important cost advantages at both the component level and completed systems levels.

We were the first to market at the 4K level, but we lost our edge to the Mostek device. The industry embarked on a path of increasing memory size that continues today. Each new generation of DRAM stores four times as many bits as does the previous one; new generations are introduced approximately every three years. The nature of the technology is such that the cost per bit of the newer generation rapidly falls below that of the preceding product. To compete in DRAMs requires that a company maintain this hectic pace of replacing the older product as soon as the advance of semiconductor technology to finer structures allows.

At the 16K level, the next generation, we introduced a three transistor per bit device. Mostek, however, introduced theirs with a single transistor cell. We moved as fast as possible to incorporate this new design, but we were too late. While the 16K was a successful product for Intel, we had lost our leadership at this generation.

In parallel with the DRAM effort we were investigating a semiconductor structure called a Charge Coupled Device, or CCD, that allowed very close packing of the memory bits. This gave a low cost per bit, but the stored information had to be addressed serially rather than at random. The lower cost structure, however, made this addressing disadvantage less important in many applications. Intel's CCD effort played a significant role in solving one of the problems with the 16K DRAM. This problem deserves special attention because it is perhaps the only example that suggests that we may be approaching fundamental limits in making practical devices smaller. Let me describe this in more detail.

AT&T, in building large memories with the Intel 4K DRAM, discovered during extensive tests that the information was occasionally corrupted, that is, some stored data was lost. This seemed to be a completely random phenomenon. But it was of considerable concern to AT&T because their computing systems did not have error correction circuitry to handle faults of this type. Any single bit error would crash the computer system. We launched a task force to investigate the problem.

I suggested to our engineers that perhaps the source of the problem was cosmic rays. An article written several years earlier by the scientists at the RCA Research Laboratory suggested that as devices approached the small size of the transistors that we were using, cosmic rays would be the ultimate limit on system reliability. In order to check out this hypothesis, we had to shield the circuits from most cosmic rays. We purchased forty tons of lead bricks and built an igloo to enclose the semiconductor memory systems under test. We soon realized that the CCD memories were much more susceptible than were the DRAMs, so our experiments switched to CCDs, where the time between dropped bits was much smaller.

The lead igloo seemed to have no effect on the rate of occurrence, but other things did. Gene Meieran, an Intel Fellow and an ardent mineral collector, showed that holding a piece of uranium ore close to the device increased the failure incidence dramatically. On the other hand, a relatively minor amount of shielding, such as a drop of epoxy on the semiconductor chip, seemed to eliminate the sensitivity. Rather than cosmic rays, the culprits causing the soft errors were alpha particles (positively charged atomic particles consisting of two protons and two neutrons emitted in radioactive decay), some of which are emitted in the natural background radioactivity in the package parts themselves. Small contaminations of such elements as thorium and uranium in the alumina ceramic packages emit minor quantities of radiation. Occasionally one of the alpha particles would impinge on the silicon where a bit was stored, causing an avalanche of hole-electron pairs that destroyed the stored data. We had located the source, now we needed a solution.

Clearly there were multiple approaches. One would be to purify the packaging material so that no radioactive elements remained as impurities. A second would be to put a barrier of material completely free of radioactive impurities between the package and the silicon, since alpha particles do not penetrate more than a few micrometers of solid material. A third would be to increase the stored charge on the silicon to the point where an alpha particle would not generate sufficient charge to destroy the information. We pursued all three without complete success. Even though the purity of the package materials has improved significantly, there will always be a residual background with which one must contend. Built-in alpha particle shields have not proven to be the solution of choice, either. To this day the limiting minimum size on DRAM cells for reliable operation is based on the charge being large enough so that alpha particles are unlikely to disturb the stored information.

Intel continued to participate in the DRAM business through the 64K and 256K generations. However, the attractiveness of the DRAM business was decreasing. Many competitors had entered at the 16K bit level, particularly several large Japanese companies in the late 1970s when the American industry was running out of capacity to supply the burgeoning demand. In 1979 at Intel, a successful 16K DRAM and the industry's only 16K EPROM were important production products. As the demand for these products grew faster than our capacity, we found that we could not satisfy our customers' demand; we were forced to make a choice. We decided that our customers would be best served if we devoted our capacity to the product for which we were the sole supplier—the 16K EPROM—and sacrificed the DRAM portion of the market where many other suppliers had products. Thus, when the Japanese entered the market, we gave them market share as fast as they could accept it.

DRAMs became the largest single category of semiconductor products, representing about a quarter of total sales. Beyond that, for a variety of reasons, success in DRAMs depended upon early use of the finest structures in the chips. Thus, they became the products that drove technology advancement in semiconductor manufacturing in the 1970s and 1980s.

The Japanese assumed the dominant position in DRAMs during the early 1980s. By the end of the strong industry recession that began in mid-1984 and lasted into 1986, they were the dominant suppliers of 64- and 256-bit DRAMs. Intel led the transition from NMOS to CMOS DRAMs with our 256K, but again at the expense of being relatively late to market; accordingly, we gained only a modest market share. We had committed to an aggressive development program to be the first with a 1 megabit DRAM, using our CMOS approach. In 1985 we were faced again with a tough decision. We had successfully developed the product and the process and were prepared to put it into production. To regain a position as a significant player in the market, we figured it was necessary to commission two silicon wafer fabrication facilities at an estimated cost of over $400 million. But looking at the DRAM business in 1985, we saw that our competitors were all losing money. The large overcapacity that had been building principally in Japan during the 1983–1984 industry boom was being kept full by dumping DRAMs below manufacturing cost on the world market. It was hard to make the $400 million commitment in a market that looked oversupplied for the foreseeable future. When Andy Grove asked me, "If you came in from outside to run Intel, would you invest in the DRAM business?," it was clear that we had to overcome our emotional commitment to the market with which Intel had originally become a successful company. We decided to abandon the largest semiconductor product type, one that we had created and that had been critical to our early success, and focus our efforts on other areas where there was a greater chance to succeed. If we could find a taker, we would sell the design and technology for the 1 megabit DRAM and concentrate on other areas.

OTHER MEMORIES

Up until 1985, the DRAM received most of the publicity and certainly most of the attention of the major semiconductor manufacturers. For Intel, however, SRAMs and especially EPROMs were far more profitable products. We evolved the EPROM progressively from the original 2048-bit chip of 1971, increasing the density twofold about every year.

EPROMs sold at considerably higher prices than DRAMs, perhaps because of our early misreading of the market's needs. When we first developed EPROMs, we priced them with the idea that they would be low volume products, used primarily for building and debugging prototypes. In fact, the advantage of erasability proved irresistible to system designers. They used them in production quantities for a wide variety of applications, but the prices were still based on the original idea. The net result was an extremely profitable product that was our major source of income for many years. We tried not to make this fact generally known and as a result had relatively little competition, especially during the 1970s when our major competitor fought for market share in the high visibility DRAM market.

EPROMs were Intel's most profitable products until mid-1984, when semiconductor prices collapsed. The price on Intel's leading EPROM fell over 90 percent in a nine-month period. A product that had been selling for over thirty dollars fell to less than three dollars. While the semiconductor industry has had marvelous success in decreasing product costs over time, this far exceeded our capability. A very attractive product was no longer a profit producer; the industry's overcapacity had spilled over to ruin the EPROM market.

To a considerable extent, this price collapse was also driven by Japanese overcapacity and dumping. Intel joined two other US companies (Advanced Microdevices and National Semiconductor) in filing an anti-dumping action against the Japanese. The US government acted with unprecedented speed and we succeeded in preserving significant US participation in what was left of the formerly lucrative EPROM market segment. A similar anti-dumping action was initiated subsequently by the US government with respect to DRAMs. Unfortunately, it was too late. Most of the US companies, including Intel, had already left that business. To reenter would have been far more difficult than it would have been to remain involved in the first place. The net result was that the DRAM dumping action did little to preserve the US industry and instead created considerable resentment among US DRAM users. They blamed the anti-dumping action and the resulting "minimum fair market prices" for DRAMs prescribed in the 1986 US-Japan

Semiconductor Trade Agreement on the agreement rather than on the market's recovery and the tightening of capacity that actually occurred in 1987.

MICROPROCESSOR EVOLUTION

Meanwhile, we continued to evolve microprocessors. The 4004 was designed to be a good calculator, but the 4-bit word length that it handled could only distinguish between sixteen characters. This was fine for decimal digits but was not enough to handle alphanumeric information in a single word. As the development of the 4004 neared completion, a major customer of Intel, Computer Terminal Corporation (CTC), which later became Datapoint, was interested in increasing the intelligence in a terminal designed principally for the banking industry. In discussing the possibility of integrating more of the function onto a single MOS device, Ted Hoff visited CTC at their San Antonio facility to see how our technology might be used to build logic for their terminal. Together Intel and CTC engineers decided that the idea embodied in the 4004 could be extended to make an 8-bit computer chip capable of handling alphanumeric information efficiently. We began to block out the architecture for such a device of roughly the same complexity as the 4004 and started the development of the 8008 for the terminal application.

CTC chose not to use the 8008, but other customers did. They used it in a variety of data-oriented controllers, such as dedicated word-processing machines. For some of the applications performance was marginal. Therefore, there was considerable demand for a higher performance product. To meet this demand, the 8080 was conceived. It would be faster than the 8008 and would remove some of the limitations of the previous device. Intel introduced it in 1974, and the 8080 became an important milestone in the development of the microprocessor market segment. It gained broad acceptance across a variety of applications, especially those related to intelligent terminals. The market for microprocessors, however, was relatively slow to develop. These components were like none that had come before. To use such complicated functions required software development and extensive debugging of both hardware and software. While they were small and inexpensive, microprocessor systems were full-blown computers and required the tools to assure they did what they were designed to do.

In order to accelerate market development, Intel found itself pulled into helping to design systems. At first, this consisted of supplying printed-circuit boards with a working design that allowed the customer to test his software. It expanded in size and sophistication until Intel was supplying complete systems for writing software and debugging both hardware and software systems. The development systems products even included the ability to program EPROMs to contain the programs. For several years, Intel's business in development systems exceeded its microprocessor revenue by a large margin. Our hope was that the development systems would eventually lead to a large microprocessor component business, but in the beginning it was the other way around.

We extended our family of microprocessors as the technology became available and the customers became more demanding. The 8080, an 8-bit processor, was extended to a software compatible 16-bit processor, the 8086, in 1978 and an 8-bit bus version of the 16-bit processor, the 8088, in 1979.

Through a large marketing effort, these 16-bit products were broadly designed into many electronic systems. We set a goal of achieving two thousand design wins in 1979, which, in fact, we exceeded. Buried among these two thousand wins was one that proved crucial to Intel and important to the electronics market overall. A small IBM group in Boca Raton, Florida, chose the 8088 for their first personal computer. While this was recognized as a significant design win, its true importance was not appreciated for some time. In fact, it changed the entire course of Intel's history.

In this application, the microprocessor was used in a new way. Previously, most microprocessors had ended up in dedicated control systems, where the function was determined at the time the system was built, and the program was stored in a permanent memory that would not generally be altered. With the PC, however, we had a user-programmable device. The program was changed at the will of the PC operator, bringing the full power of the digital computer to the individual. This had been tried before. In fact, Apple Computer was developing a successful business based on a similar model, but they based it on a competitor's microprocessor. For Intel, the IBM win was crucial. With the PC we began the spiral of ever-increasing software sophistication followed by increased demands for processor capability. We have been riding that spiral ever since.

The early 1980s, with IBM's introduction of the PC-AT, saw an upgrade of the 8086 to the 80286 for PC applications and other high performance data processing requirements. In 1983 the entire electronics industry was booming. Intel's new product families, including the 80286, the 80186 (a chip that included more than just the processor), and our new line of microcontrollers (single chip computers dedicated to specific application functions, i.e., automobile engine control or control of a VCR), were growing rapidly in production volume. All of these parts hit very receptive markets, and our customers did not want to rely on a single supplier to support their volume requirements. After all, if a microprocessor was not available, they might not be able to ship a computer. Consequently, our customers would limit commitments to our products unless we obtained alternative sources to provide virtually identical products. A half dozen sources for the microcontrollers were put in place, and the 80286 was given to American, Japanese, and European competitors. The boom continued until mid-1984, when demand slowed dramatically. It turned out that our customers needed only one-third of the chips that they had predicted earlier, and we could have built them all ourselves. We were now in a quandary common to the semiconductor industry. There was excess capacity across the industry. Prices collapsed and the industry rushed into its 1985–1986 depression. At Intel, from 1984 to 1986, our staff dropped by more than eight thousand, a reduction of over 30 percent of our work force. We were thrown into a loss position in 1986, the first time since becoming a public corporation in 1971. Intel was not alone. The semiconductor industry worldwide lost several billion dollars during 1985 and 1986.

One thing that has always been true in this industry is that recovery after a down cycle never occurs with old products. Technology evolves so rapidly that the market moves to the next generation or beyond. Thus, to be successful it is necessary to continue investing in new products even during these down periods. Intel did this. In particular, we saw that it was time for another major expansion in the 8086 compatible architecture. The world was asking for processors with full 32-bit capability. While it was not obvious how to extend the 80286 architecture, a group of engineers led by John Crawford saw that they could preserve software compatibility and eliminate many of the problems of the earlier 8086 and 80286 architectures. To do so would require a significantly more complex chip than had been designed previously. While the 80286 had some 130,000 transistors, the 32-bit successor, the Intel386™ chip, would have 275,000. The expected advantages included much higher performance and an architecture where software could be written to operate with greater reliability and security. This, again, was an extremely important product in Intel's history and probably the foundation of our success over the last decade. Most members of the executive staff remembered the poor experiences that Intel had with the i286 second sources. We wanted to obtain equivalent value in exchange for a i386 second source. Intel and Advanced Microdevices (AMD) had agreed that each company would earn points that could be used to evaluate technology exchanges. But by 1987, there were no acceptable AMD products for exchange, and nothing was expected for at least two years; AMD had no way to earn second source rights to the 386. We decided that we would not transfer the 386 to AMD unless we believed that we were actually going to get valuable products in return. Thus, there was no second source arrangement with any 386 competitor other than one that allowed IBM to make a portion of its internal consumption. Intel management was prepared to go it alone without an alternate source. We dealt with our customers' desire for alternate sources by undertaking a large capital expansion program to build this complex chip in multiple, independent Intel factories.

This was a real change in the way that the industry operated, but because of the carryover of the software from previous generations, we thought that there was sufficient reason for customers to continue to use the product, even with a single supplier. Our major customer at this time, IBM, questioned our ability to make a product of the complexity of the Intel386 processor and decided to stick with the 16-bit i286 for one more generation of PCs. When we were successful in developing the chip, they had no programs to take advantage of it. Compaq Computer, on the other hand, took the risk that the Intel386 microprocessor development would be successful and devoted a major effort to developing a family of systems based on the

Intel386 chip. When they introduced these in 1986, they assumed the role of technical leader in the personal computer market. Their product, the Deskpro 386, set a standard that other players in the industry were forced to meet.

The Intel386 chip was highly successful and kept us busy for several years as we tried to meet the market demand. This has been followed by the Intel486™ microprocessor family and today the Pentium® and Pentium® Pro processors. The market for PCs is approaching the $300 billion mark, and Intel's business has become strongly focused on desktop computing and associated interconnection and enhancement products. Essentially, we have made the transition from a semiconductor company to a computer building block supplier based primarily on semiconductor technology.

Many of today's chips, however, with their thousands of times increased complexity, sell for no more than did the first Intel products. The past twenty-eight years have seen wild swings in the market. Shortly after we were founded, we entered a recession period, something that actually can be advantageous for a start-up company because people and materials become more available. It is, however, a difficult time in which to grow. This was followed by a boom in the early 1970s, which ended with the 1974–1975 oil embargo crash. From 1976 through 1980, Intel and the industry enjoyed a long uninterrupted period of growth, which was terminated by another recession in the early 1980s. Intel passed the billion dollar revenue mark in the subsequent recovery, peaking in 1984 prior to the collapse in 1985. Recovery from the trough has been steady since 1986, with Intel's annual revenue in 1995 exceeding $16 billion. It has been an exciting twenty-eight years. Technology has progressed dramatically. In the beginning we were pushing the state of the art to put two thousand transistors in a chip. We now put several thousand times that many in a modern microprocessor and envision that this rapid evolution should continue.

George N. Hatsopoulos

A Perpetual Idea Machine

STOCK ANALYST RECENTLY described Thermo Electron as *a perpetual idea machine*. I started the company in 1956 with only $50,000. Today, Thermo Electron is a world leader in environmental monitoring and analysis instruments and a major producer of paper-recycling equipment, biomedical products, including heart-assist devices and mammography systems, alternative-energy systems, and other products and services related to environmental quality, health, and safety. Although Thermo Electron's annual revenues are now well beyond the billion dollar mark, the company remains as it started out—a company dedicated to entrepreneurship based on technology.

Thermo Electron is actually a family of publicly-traded companies, the result of our strategy of "spinning out" promising businesses to bring innovative technologies to commercial markets. We found that this unusual approach affords an ideal climate for sustained growth. It allows us to raise the necessary capital for our diverse new ventures and provides a focus for the creative energies of individuals seeking to apply their ideas to the emerging needs of society. Over the past decade some of our spinout "offspring" have established spinout companies of their own, which we fondly refer to as our "grandchildren."

The evolution of Thermo Electron is perhaps the reflection of an idealized vision of my own life. In my "ideal life" I would first study physics in order to understand its fundamental laws. Then I would make a significant discovery that would lead to an invention, use that invention to start an enterprise, and build an organi-

George N. Hatsopoulos is the Founder, President, and Chairman of the Board of Thermo Electron Corporation.

103

zation that would become the world's largest technology-oriented corporation. And I would accomplish all this in, say, fifty years. Of course, Thermo Electron is not yet the world's largest technology company, but neither has it been fifty years.

I was born in Greece in 1927. Both my grandfathers were politicians, members of the Greek Parliament, and my father was the chief operating officer of the country's electric rail system. As far back as I can remember, I was fascinated with machines. At the age of seven, I "invented" an electric iron, hoping to replace the kind that my mother heated on the stove. Of course, I was disappointed to learn that someone else had already done this twenty years earlier. As a teenager during the Occupation of Greece in 1941, I spent a lot of time in the library reading about other countries. One thing that appealed greatly to me was America's unique quality of entrepreneurship. One of my idols was Thomas Edison. Although he had started General Electric, he lost control of it because of his poor business skills. I began to think seriously about becoming an inventor who would also be a good businessman. I do not pretend today to be as good an inventor as Edison, but I think that I have become a better businessman.

My first experience in combining technology with business came about as a result of the Occupation. The Germans manipulated our radios so that the only station we could receive was the one they used to broadcast their propaganda. They sealed the radios so that the setting could not be changed and inspected them every month; if the seal was tampered with, you were arrested and then sent to a concentration camp. And so the Germans unknowingly created a market for radios that were not sealed.

I started making radios in the basement of our house, giving transmitters to the underground and selling receivers to anyone who dared to buy them. My father did not find out about my "business" until after the war, and even then he nearly had a heart attack after learning about it. I do not consider the Occupation to be the worst years of my life even though many people, including my family, went hungry and lived in constant fear (dead bodies were found daily on our streets). Despite the hardships and the terror, there were some positive elements of the Occupation. One that really stands out is the way people came together. Because we all were "in the same boat," there was no longer any separation of classes by education, wealth, or social position.

Those were exciting times, despite the risks, and much that I learned has stayed with me. I think that perhaps the most important thing I learned was to ask myself, what is really going to be *needed* in this society? I knew that when the need is severe enough and arrives unexpectedly, there will be a great opportunity for someone to start a new enterprise. I believe that this is especially true with technology-based ventures. When the war ended, however, I learned another valuable lesson: even the most exciting markets can suddenly disappear. Similarly, even the hottest technologies will eventually be supplanted by more innovative ones.

After the war, I entered the Athens Polytechnic, which my father had attended and where some of my uncles served on the faculty. My original intent was to study electrical engineering, but I became intrigued with thermodynamics. Because I was unable to obtain answers to my questions about certain thermodynamic principles and theories, I decided in 1948 to come to the United States and study at the Massachusetts Institute of Technology (MIT), where I received both my BS and MS degrees. After serving in the US Army, I returned to MIT to earn my doctorate in mechanical engineering, and in 1956 I was appointed an assistant professor.

However, it was still my dream to build an enterprise based on innovative technology. There is tremendous satisfaction in creating something that bears one's own personal imprint. That is why many of us have children, write books, make inventions, develop new scientific theories, and, of course, start our own companies. An important motivation for me is the competition that is involved. I really only enjoy playing tennis when I have a chance to beat my opponent. Competition, especially that which is inherent to business, is something from which I derive great satisfaction. There are people who are averse to risk, but I am not, provided the potential gains are worthwhile. Even if the probability of success for a particular new venture is small, it makes sense to pursue it if the rewards seem big enough. My idea for establishing a new company was to create a broad-based technology enterprise that would work simultaneously on a spectrum of promising innovations, each of which might involve significant risk. Of course, such a collection of fairly risky ventures has considerably less overall risk than does each of its parts.

In April of 1955, I joined my friend, Peter Nomikos, for dinner at the Harvard Business School, from which he was about to graduate. (I often ate dinner with him there; the food was better than at MIT.) For my doctoral thesis at MIT, I had invented a thermionic converter known as a thermo-electron engine, a compact and effective device for converting heat directly into electricity without any moving parts. At dinner that evening, Peter and I spoke at length about our visions for the future. Eventually, we conceived the idea of forming a company to exploit my invention. My late uncle, Costas Platsis, and Peter's father were bridge partners, so we convinced Costas to lobby Peter's father on our behalf to raise some start-up money for us. In our initial business plan, we asked for $50,000 to build a prototype and achieve proof of principle. The money was granted with an ease that should have given us pause at the time. We learned much later, after Thermo Electron was already a huge success, that upon reading our proposal, and knowing nothing about thermionics, Peter's father had said to Costas: "Listen, Costas, I will give them the $50,000. They will lose it. Then George will go back to teaching at MIT and my son will concentrate on the family business. They will both benefit, and it will be a small price to pay for an important lesson."

The first thing I did was to build a working thermionic converter. Unfortunately, we have yet to find a commercial application for the device. Nevertheless, our research and development experience allowed us to pursue other ventures. We found clients who needed our knowledge of thermodynamics, who were willing to fund us with research contracts. Initially, our main customers were agencies of the federal government and some gas utility companies. To expand our market, we looked for emerging problems in society that might be solved through commercial applications of our growing expertise in thermodynamics and related technologies. We began developing products that could serve such markets and also bring us a healthy return on our R&D investments. Many of our people worked long hours to develop ideas that later became Thermo Electron product lines. In fact, some of our most important lines of business today grew out of those projects initiated in the early 1960s.

By 1966, our annual revenues had grown to about $2 million, and the following year we went public to raise capital for continued R&D and to set up operations for commercializing our first products. We also started acquiring companies that could speed the process of converting our technologies into products and bringing them to market. Some of our research contracts with gas utilities required that we commercialize the technology by a certain time or forfeit our patents and rights. Buying a company with established production facilities and marketing channels was often the only way to meet the deadline.

An example of how we tried to match the company's development efforts to meet the specific needs of society can be seen in our response to the Clean Air Act of 1970. That legislation required automobile manufacturers to limit the nitrogen oxide (NOx) emissions of their vehicles. The problem was that no practical instrument for measuring such pollutants existed at the time. We had been working on an automotive steam engine for the Ford Motor Company and were trying to convince Ford's engineers that our system had very low NOx emissions. In order to confirm our claims, Ford sent two of their employees to measure our engine's exhaust with a laboratory model of a new NOx instrument that could give instantaneous results. After seeing that prototype in action, we decided that it was something that everyone was going to need. When we asked about Ford's plan for this exciting technology, they indicated that they had offered it to the Beckman Corporation, then the world's leader in instruments, but Beckman believed that it would take two years to produce the new device. I told the people at Ford, "Give us an order and we will produce the instruments in three months." They thought I was crazy, but no one else had a device to meet their needs, so they gave us an order for twelve instruments. We succeeded in delivering the instruments in ninety days, and although these first NOx detectors were not perfect, they were the first on the scene. Shortly thereafter, the Environmental Protection Agency designated our technology as the standard to be used for all NOx measurements on automobile exhausts. Suddenly, automobile manufacturers from around the world sought us out. We did not need a marketing staff—just a telephone operator to take the orders.

The next significant opportunity for Thermo Electron arose as a result of the energy crisis triggered by OPEC's 1973 oil embargo. This event caused market dislocations of global proportions, creating a new set of market needs. Prior to the embargo, we had undertaken some thermodynamic studies to determine the efficiency of energy use in various industries. We found that industry was using more energy than any other economic sector—about 40 percent of the total US consumption. Our studies showed that the industrial use of energy in the United States was less efficient than that of any other major industrial country. It was clear to us that enormous savings could be realized by improving industrial processes through the application of modern technology.

About a year before the oil embargo, we were asked by the Ford Foundation to assist them in preparing a comprehensive study of energy use in the United States. They needed our input on the industrial sector to complete their Energy Policy Project. We selected for our study six industries that accounted for roughly 40 percent of industrial energy consumption. Based upon rigorous thermodynamic principles, our investigation pinpointed losses and defined the true efficiencies of various industrial processes. This study became a landmark in the energy debate of the 1970s because it was the first time anyone had really looked at industry in sufficient detail to establish absolute (i.e., Second Law) efficiencies. The conventional wisdom at that time held that most industries were fairly efficient. The petroleum refining industry, for example, was said to be 92 percent efficient, whereas our analysis proved it to be only about 8 percent efficient. The problem was that no one was properly applying the complete set of thermodynamic laws when determining the efficiency rate. As a result, they were not seeing the great potential for energy savings that exists in so many commonplace processes and devices.

Although we had initiated our study as a means of gaining a basic understanding of industry's energy practices, the timing could not have been more perfect. Our findings were published just weeks after the 1973 oil embargo, when energy prices were rising dramatically and everybody was struggling to cope with this new problem. The company's growth in the 1970s was due in large part to our development of energy-efficient equipment for industries such as papermaking and metallurgical processing.

We continued to study the issues surrounding energy use and in 1976 published results showing that the United States could sustain its growth in GNP for many years without increasing overall energy consumption. This is in fact what actually happened, although at the time we were called "unrealistic academics." Our findings did, however, gain the attention of policymakers and legislators, and we contributed to the drafting of the Public Utilities Regulatory Policy Act of 1978. This important legislation opened the door to competition and deregulation in the electric power industry, which then facilitated the introduction of cogeneration and other energy-efficient technologies.

Thermo Electron grew rapidly in the middle and late 1970s as we introduced products such as heat-recovery recuperators for industry and cogeneration equipment for the efficient production of both electricity and process heat from the same fuel source. A massive shift in investment priorities had taken place as Americans began exploiting opportunities for getting the most out of their energy consumption through improved equipment and processes. Although it was clear that energy efficiency was improving throughout the economy, we could see that only a small fraction of the possible energy-saving measures were actually being implemented. That led us to believe that something was inhibiting the application of new technologies. We soon learned that the problem was due primarily to America's low cost of energy relative to the cost of capital.

For years, America's managers had responded in a perfectly rational way to the country's lower energy prices by foregoing capital investments in energy-efficient processes. They were no different from homeowners who chose not to insulate their houses or buy efficient refrigerators or air conditioners that had high initial costs. If the price of energy is high enough compared to the cost of capital, then, of course, it is advantageous to buy energy-efficient technology. In effect, you can substitute capital for energy. The Carter administration, unfortunately, never understood this. They thought that the way to conserve energy was to do without—in other words, lower the thermostat. We countered, "No, we want the thermostat to be up, but we want more efficient use of energy. Instead of energy *conservation* we want energy *productivity*—in other words, have your cake and eat it too."

Energy was only one part of a larger issue affecting the American economy; another was the decline of our competitive position in world markets. Throughout the 1970s, there was a significant difference between the cost of capital in the United States and in Japan. In 1982, I embarked on a study of this differential, which revealed two main factors. First, the interest rates in Japan were much lower than those in the United States because of government policy; and second, Japanese companies were very highly leveraged. Since the interest on debt is tax-deductible, whereas the return on equity is taxable, there is a big advantage in overall cost-of-capital in leveraging with high debt. This led me to my study, *High Cost of Capital: Handicap of American Industry,* which attracted considerable national attention. Until that time, people who should have known better did not really understand what constituted the cost of capital. It was actually a complicated mixture of equity, debt, and leverage. Moreover, the equity price is not easy to compute, and it is affected by tax policy. By applying rigorous scientific methods, we were able to calculate more accurately the current and historical values for cost of capital in different countries.

After our findings were published, many other studies were conducted on the subject. Once we understood the real mechanisms affecting the cost of capital, it became clear how policy should be altered. So we repeated what we had done with the energy policy: we studied the fundamentals, came up with proposals that made sense, and then tried to influence national policy.

In developing our business plans for the 1980s, we expected to make major investments in energy conservation technologies. It was assumed that the price of oil would continue to rise for many years; there were even some predictions that by 1990 the price of oil would be $40 to $60 a barrel. In this environment, we planned to focus on developing and producing energy-efficient equipment. But by 1982, there was a major recession underway, and it hit hardest in the basic industries—from automobiles to steel to farm equipment. It created the rust belt. These were the industries that we had expected to be our prime customers for energy-efficient process equipment. At the same time, there was a glut of oil. Energy prices dropped dramatically, pulling the rug out from under industry's incentives to buy more efficient equipment. Almost overnight we found ourselves without a market—our customer base had simply disappeared.

Fortunately, we had simultaneously been working on a number of new ventures. More development was needed, however, before they could be marketed, and we had planned to invest our earnings in that development. This is how we operated—we always had new technologies in the pipeline so we would be able to create additional business ventures. The problem was that in 1982 our earnings dropped by a factor of three. We were suddenly faced with a tough choice: either stop development of any new ventures in order to economize, or wipe out our earnings

entirely to pay for accelerated development of the new businesses. We chose the latter, which brought our earnings in 1983
to near zero. In order to minimize the impact of this decision on
the price of our stock, we explained our situation to several Wall
Street analysts. We told them that if we did nothing, our earnings would drop by 60 percent. If we economized, the drop
might be limited to 40 percent. We explained, however, that we
were going to spend all of our earnings to speed the development
of new businesses. (Fortunately, Thermo Electron has always
maintained enough cash to be able to buy our own stock at very
low prices and sell later when the price recovers.)

We were at a crossroads in more ways than one. One of our
concerns in the late 1970s, when we first created separate divisions,
was how to reward the various entrepreneurial groups that were
being formed. We tried paying cash bonuses, but basing rewards on
short-term financial results cannot reflect the true value of what
has been created. We decided that the most effective way of giving
our managers incentives was through stock options. But giving
options for the parent company to each of those entrepreneurial
groups meant that the rewards would be the same regardless of
whether any particular group was successful or not. Such options
were not tied to specific achievements. So we decided that if we
could have a stock price for each of the divisions, we would be able
to develop our new enterprises using the same device that had
originally built Thermo Electron.

We first put this theory to work by spinning out our artificial
heart venture. The company had worked for several years on
artificial heart research, with over $70 million of funding provided
by the National Institutes of Health. (That is one of the advantages
of working on socially important technologies—the government is
often willing to subsidize their development.) In 1982, the government drastically cut funding for heart research, and we did not
think that we would have a viable product for at least another
decade. We needed to raise more capital to support further research
on the artificial heart, but we were not sure that investors and
analysts who followed the company's stock would understand the
heart's potential. They often would say, "I hear you are working on
artificial hearts. What does that have to do with energy?"

For these reasons we decided to sell stock in a newly created
public subsidiary, Thermedics, Inc. I worried about selling off a

core business, but Thermedics was small compared to the whole of Thermo Electron, and so in June 1983, we sold eighty thousand shares of Thermedics to a private venture capital firm for $8.00 a share. In August, we offered seven hundred thousand shares to the public at $9.50 a share. Thermo Electron retained 84 percent of Thermedics.

I prefer the word "spinout" to describe what we have been doing because the new companies are no longer divisions or subsidiaries, nor are they "spin-offs," which to me implies companies in which you sell your interest.[1] Spinning out Thermedics not only created a new, smaller company, but also renewed Thermo Electron itself by launching a new venture rather than letting us rest on our laurels. And it gave our managers a chance to be entrepreneurs. We had found the way to give our top people a stake in the business that would directly reflect the value each one added to the company. It is better to let the market determine what value has been created because it provides a highly objective criterion based on real-world conditions. Also, there is a tremendous advantage in having entrepreneurs deal directly with shareholders.

There was also an unexpected dividend for Thermo Electron as a result of spinning out Thermedics: now that this venture stood alone, investors could see its value as a "pure play." And Thermo Electron was valued even more since it was the majority stockholder in Thermedics. Thermo Electron stock went up, and this was a benefit of the spinout that we had not anticipated. Spinouts became the model for our development as a company throughout the 1980s and up to the present. Whenever a new venture reached a certain point of maturity—typically, when it had a strong management team with a good new idea and the capacity to grow at a compound annual rate of about 30 percent over a long period of time—we would spin it out as a publicly-owned company in which we retained the majority interest.

We even spun out the company's R&D core, which became ThermoTrex. We are now in the process of rebuilding the R&D center while managing ten highly successful public spinouts in which we hold majority stakes; several privately-held, majority-owned spinouts; and seven wholly owned subsidiaries. Currently, our biggest unit, Thermo Instrument Systems, with about $700 million in annual sales, is thinking very seriously of establishing spinouts of its own. I am encouraging them to go ahead.

The only way to obtain capital at a reasonable price in this country is to focus on the entrepreneurial inclinations of certain investors, people who are not interested in a huge conglomerate because they cannot get a feel for anything it does. We repackaged our equities so that investors can buy a piece of any promising technology that excites them. Thermo Electron has gone much further toward decentralization than any company I know. I think that setting up many of our divisions as publicly-owned organizations gives us a highly flexible and responsive structure. Managers of each subsidiary know that their actions will be scrutinized not only by corporate management but also by their own shareholders.

I firmly believe in the benefits of small enterprises. However, small companies standing alone have some big disadvantages—they usually do not have strong financial support, they often lack management know-how, and they do not have market leverage. At Thermo Electron, we have connected a set of small companies into a family that provides financial resources, management support, and strategic direction. At the same time, our companies are able to act independently and respond to the needs of their own customers and shareholders. It is the best of both worlds.

Our companies are autonomous, yet together they actually comprise the larger entity that is Thermo Electron. There is a core in the sense that our experience and our strong finances originate from the parent company, which can raise capital and consolidate legal and administrative services. We also generate new ideas there and manage businesses that have yet to be spun out. The strategy is central, the financial resources are central, and all the administrative services—from the legal department to the human resources department—are central. The spinouts are implementors of whatever strategy is agreed upon by the parent company.

Thermo Electron is now more like an American *keiretsu*[2] than a conglomerate. This is not a coincidence. In studying the advantage that the Japanese have in the cost of capital, I attributed a good deal of that advantage to their *keiretsu* system. When I first presented the spinout strategy to our board, I said, "What we want to do is to build an American *keiretsu*." Still, there are differences. Our companies are a lot closer than the companies in a *keiretsu*. We transfer employees from one company to another when it is agreed upon by both sides, and there is much more interaction among our companies. We share the same kinds of technologies—

energy and instrumentation are the primary foundations upon which virtually all of our businesses have been built.

I do not think that any other large American companies have yet mastered the ability to create new businesses. There are companies that add or buy new businesses, but they do not create them from within. I think our approach represents a way for American companies to become incubators of new business. Traditionally, most new businesses are generated by venture capital, by individuals. That is how *we* started, along with Digital, Polaroid, Data General, Lotus, Apple, and even Hewlett Packard. All of these companies began as new ventures, where a group of entrepreneurs got together to launch some start-up companies. But none of them were able to duplicate their original success after they grew. Thermo Electron has developed a technique that sustains the entrepreneurial culture. Our growth is driven by inventing new technologies as the need arises, and we do not get into ventures where we do not understand the technology. Our structure provides one-stop shopping for entrepreneurs. Nobody ever has to leave Thermo Electron to become a great start-up success; they can start their own companies right here.

Thermo Electron's success story probably could not have happened anywhere but in America. I think this country has tremendous competitive advantages but they are not being fully utilized. Many great innovations, for example, originated in the United States but were first commercialized overseas. In the past, Japan has often picked up on our breakthroughs and exploited them in world markets. America's primary advantage, I believe, is the entrepreneurial spirit that pervades our society. Being an entrepreneur is part of the American dream, and that is why we are the only country that has continually created new businesses and industries from scratch. In Massachusetts, for example, the majority of the state's successful companies were started after World War II. In Europe, by contrast, today's top-ranking companies are the same ones that dominated industry before the war. Start-up businesses in America derive great benefit from this country's appetite for exciting new ventures. Venture capital for risky investments is much more readily available here than in other countries. Moreover, our vigorous stock market affords the opportunity to recoup successful investments through initial public stock offerings. The same kind of financial market flexibility and diversity is not available to newly created businesses overseas.

A key question facing America's managers is how to preserve an entrepreneurial climate as companies grow larger. At Thermo Electron, we have found that the spinout approach provides at least part of the answer. Perhaps other companies will wish to adopt similar strategies. I am convinced that only those companies that can continually renew themselves by spawning new businesses will have a chance to go on indefinitely. There are still questions to be answered, and Thermo Electron should be viewed as a "work in progress." We are able to hold our family of companies together now because we all grew up together. Few people who start their careers at Thermo Electron go elsewhere. How will the next generation keep our growing number of businesses working closely after this group retires? When the company becomes very large— say ten times today's size—will they still be able to strike an effective balance between autonomy of operations and close coordination of major strategic decisions? The challenge, not only for Thermo Electron but for American industry at large, is learning how best to capitalize on this country's unique competitive advantages. If we can do this, I am confident that the United States will remain the industrial powerhouse that it has always been.

ENDNOTES

[1] Thermo Electron has never sold any shares of its subsidiaries. These subsidiaries sell shares to the public to raise capital for their own use. As a result, the parent's ownership always falls below 100 percent.

[2] *Keiretsu* is a group of companies federated around a major bank, trading company, or large industrial firm. Affiliation generally implies extensive reciprocal ownership of common stock within the group. Accompanying these cross-shareholdings are implicit but widely understood and rigorously observed mutual agreements not to sell shares that are held reciprocally. Many of the companies have common directors on their boards. See W. Carl Kester, *Japanese Takeovers: The Global Contest for Corporate Control* (Boston, Mass.: Harvard Business School Press, 1991), 54–57.

William M. Haney, III

The Power of Invention

T HE PAST TWENTY YEARS have seen profound changes in the structure and relationships of the international community. The Soviet Union has collapsed, China's economic structure has been transformed, and a move toward democracy is sweeping the developing world. Population expansion and broadening industrialization continue to put painful pressure on the natural world while demographic and technological change of an exhilarating scale are redefining the characteristics of and opportunities for the work force worldwide. As a nation that enjoyed historically unparalleled economic superiority only two score years ago, it should come as no surprise that many Americans find these changes offer our nation little benefit and much risk. Even as we stand as the world's sole superpower, many point to the gains that other countries' economies have made relative to the United States' over the past two decades as evidence that our strength is on the wane, that the present nexus of geopolitical restructuring and industrial revolution bodes ill for America, and that the canary in the coal mine is the vibrancy of our domestic economy.

The decline of American domination in the world's economy does not trouble me for several reasons. First, it was to be expected that the economic position that the United States held fifty years ago would diminish in relative terms as other economies rebuilt from the devastation of World War II. Indeed, we nurtured their growth as part of an international campaign to construct an alliance of democracies to contain the Soviet Union during the Cold War. Interestingly, our present percentage of

William M. Haney, III, is President and Chief Executive Officer of Molten Metal Technology, Inc.

117

global production does not compare unfavorably with our average status over the past one hundred years but merely with the unique position that we held fifty years ago. In fact, the American economy remains the world's largest and the American worker is the world's most productive.

Second, I believe that the dislocations in our economy that are disquieting to many are actually signs of our nation assimilating the staggering technological advances of the past twenty years. The fundamentally decentralized nature of our social system has allowed us to react to the new industrial environment more rapidly than our competitors. In fields that are the vanguard of the transforming economy, from biotechnology to advanced agriculture, from supercomputers to entertainment supermarkets, we continue to lead the world as we do in forestry products, coal mining, and chemical manufacturing. The American automobile industry is increasing its domestic market share and the world's most productive steel plant is now found in the lower forty-eight states. We have accepted vast changes in the size and characteristics of our work force, a profound movement in the source of job creation in our economy, and a fantastic shift in the pace of international technology transfer. Unemployment remains low and adjusted per capita income remains high by comparison with most of our economic competitors. We continue to maintain an open and free political system and to provide our allies and economic challengers a defense umbrella that protects democratic societies throughout the world. Even now, the foundations of those nations are shuddering as they struggle to remain politically stable and economically dynamic under the same pressures that we feel at home. Countries around the world are converting to our economic model, and the fundamentals of our political system attract admirers in record numbers as they vote with their feet and seek to immigrate into our great land. Thus, while I do believe that the United States is at another crossroads in its development, both as a nation and a source of ideological inspiration to others, my sense is that it is in control of the wheel.

Despite my optimism, I do see trends that concern me as our nation wrestles to maintain its commitment to the values that have long sustained it. Fiscal sobriety seems in painful shortage in both our government and our households. There has been a disturbing erosion in the level of education of the average American worker and a corresponding reduction of confidence in our public school systems. The size of government has expanded in a way matched only by its increasing role in our society, and many voters believe that corruption is endemic in our political system. Our national self-confidence is eroded by a daily dose of violent stories, while revealing pain and defrocking heroism is an increasing part of the media's contribution to our society. Moreover, a combination of economic pressure and shifting social mores has resulted in an alarming decrease in the percentage of American children raised in stable and nurturing households. In this truly fundamental way, our future is at risk.

These trends drive my concern about the crossroads at which America stands. I believe that the faith that each American has an opportunity to build a future that is brighter than his past and to grasp a chance for principled and creative self-expression within a dynamic capitalist economy is the ideological foundation upon which the social compact that unites all the diversity of America rests. In the tumult of change we cannot afford to lose our commitment to the values that animated the creation of our country. We need to remain committed to providing effective public education as a universal good. We must celebrate diversity as a wellspring of cultural and economic dynamism, extending a generous welcome to immigrants pledged to the principles of our democracy, balancing the relationship between rights and responsibilities, practicing fiscal sobriety, and maintaining an unstinting dedication to a society where with labor and luck, insight and passion, an American can fulfill his dream of opportunity. These values provide the cornerstones upon which a society of excellence has been built and can be sustained. They provide the bedrock upon which confidence in the future stands. A weakened commitment to them, and a concomitant diminishment in our willingness to stride boldly and confidently into the future, to face the painful creative destruction that a dynamic capitalist economy demands, and to nurture and prize the pioneer spirit to create, will allow the bureaucratic need to control and justify to condemn us to building a self-limiting proposition of economic mediocrity and social turmoil. In short, providing the social conditions that foster a reasonable belief in the American dream is a critical requirement to seeing the American economy successfully through the present crossroads and confidently into the promise with which the future beckons.

I believe that the American dream is vibrant and thriving, despite the shrouds of doubt we have begun to hang upon it. In part this conviction comes from my sense that the somewhat chaotic individualism of our entrepreneurial society, with its doubts about hierarchy, is well-suited to today's fast-paced and global world. But, more substantively, my convictions are bred in my experiences. Born into a small family in a small town in America's smallest state, I have been accorded an extraordinary opportunity to pursue my goal of developing technology to address the world's environmental needs. Neither a scholar nor an intellectual, I provide a simple account of my background and experiences in the hope that it will contribute to the discussion of how American industry successfully bridges those obstacles that confront it today.

The more I read of sociologists' fascination with the dysfunction of the American family, the more grateful I am to my parents for providing me a childhood of loving care and simple contentment. Nestled on five hundred acres of rolling country along the shore of Narragansett Bay in Portsmouth, Rhode Island, is the Catholic boarding school where my father taught chemistry for twenty-three years and where I grew up. A small community unto itself, the school is built around a Benedictine abbey and the monks who populate it. Lying within what was then a small farming community, the physical assets of the site were tremendous. Tennis courts, football and baseball fields, boat houses, movie theaters, art studios, music halls, and glorious libraries, which stood as a stark reminder that love of learning was all that challenged the works of God in the hierarchy of values to which the school was committed, all these lay within the shadow cast by the church's spire. Even more exciting to me, however, were the hundreds of students and faculty members who spent at least part of the year in residence, working together through the struggle and awakening of adolescence and providing ample playmates and role models to a child.

Two teachers with a young family, my parents were drawn to this stable community because they shared its love for learning and relished the pleasure of both educating and being educated. In this quietly diverse community, my siblings and I were taught affection for the written word, the joys that simple communication can provide, the sustenance of a gentle relationship with God, and the confidence of a clear set of personal values. We were taught to honor all men and women for who they are and what they stand for, not for what they have. Our commitment to achievement was built, our interest in independence was honored, and we were introduced to the benefits of striving for excellence. We were allowed to experiment often, and, in the midst of failure and struggle, we never failed to find love. Despite the headaches that our youthful quest for freedom and iconoclastic expression caused and the stinging rebukes that its clumsy and boorish manifestations merited, our individuality was honored. We were left with the willingness and confidence to act. We did not fear to fail.

We never traveled to foreign lands, hiked the Rocky Mountains, attended the theater in New York, or spent time with America's mighty. There were no businessmen of position in our town, no great explorers, nationally known musicians, or famous folk of any complexion. We were not allowed to watch much television, and we attended the town's public schools until we were old enough to attend the Abbey. We seldom went to a restaurant, but listened passionately to the Red Sox on the radio and made weekly pilgrimages to the local libraries. It was a peaceful life despite my siblings' and my desire to make it otherwise. My first mentors were people whom I think fondly of today—a retired tent maker who had become a taxi driver, the warmhearted Portuguese-American handyman who acted as the school's plumber, a garrulous Irish auctioneer, and an extraordinarily eccentric monk-mathematician—they all figure prominently among my childhood memories. These are the people who helped my parents set our values. I remember them more vividly than I remember most of the teachers who were titularly responsible for my education, perhaps because they gave me what a young boy cherishes, namely, unrequired attention and studied respect. Much as my parents, they laughed often and deeply, cherished playfulness, and found joy easily.

At an early age, I discovered that formal education was but one way to learn. My parents' love of reading exposed me early on to the romance, mystery, fantasy, and studied exhilaration that was to be found in books and the citadels that housed them, public libraries. When licked by trouble in school or out, I turned to books for warmth, excitement, and escape. It was here that I developed such admiration for technology entrepreneurs that they stood behind only ice hockey players in my views of achievement. They seemed passionate, creative, constructive, and liberated; in many ways the history of our land seemed to be the history of these fascinating people. Remington, Franklin, Whitney, Kettering, Ford, Edison—from their minds seemed to spring ideas that had transformed American society, and with their deeds they forged a future for many, including themselves. My father, worn by the enervating petty politics of a small institution and frustrated by the lack of freedom for creative expression that his job accorded, encouraged me to work for myself, and so I dreamed of being self-reliant.

At seven years old I had a lemonade stand positioned strategically across the street from where tour buses dropped visitors touring local gardens. I remember little of achievement but much of pleasure. By nine years old, I had my first paper route. A source of ready cash and of no small stature, it accorded me a simple introduction to the working world. While I gave it up four years later to attend the Abbey, it was the first in a series of jobs, the kind that fill the childhood of many Americans—greeting card salesman, caddie, carpenter's assistant, plumber's helper, busboy, and waiter. All a predictable mixture of limitation and lassitude, for me they were but stepping-stones for my favorite work experience through high school.

Erecting circus tents for horse shows, fairs, weddings, and parties was the work of the Newport Tent Company. Founded by a thoroughly charming, intense, and generous-spirited man, the tent company employed dozens of children, mostly Irish-American and largely from working-class families, between the ages of fifteen and twenty-five. The company served as a bridge from adolescence to adulthood for many. The spirit of its creator, a fun-loving local figure who had traipsed from college to college across the east coast trading upon his intellect and athleticism until his unwillingness to conform forced him to move once more, filled the trainees. We were given both real responsibility and great freedom as we traveled throughout the Northeast, putting up tents, drinking beer, and learning from one another. It was hard work, it was liberating work, and I enjoyed it thoroughly. In a rough and ready way, it prepared me well for the years that followed.

Harvard College is an extraordinary place at which to study, but it is also impressively expensive. Despite my parents contributing what was a great sum for them, combined with Harvard's generous offer of scholarship support and my summer savings, I was left with a shortfall in my ability to pay for school, a gap I intended to fill by working throughout the year cleaning the school's dormitories. I soon had a problem, however. Earning four dollars an hour scrubbing other students' bathrooms early each morning not only strained the biological clock of an eighteen year old but also proved unsuccessful in providing the money that I required.

After several months of this labor, a fellow student and I had an idea that would release us from these constrictions. We had noticed that although many of the dorm rooms came equipped with fireplaces, firewood, due to the school's urban location, was not easily obtainable. Over a period of several weeks, we convinced a number of students to place orders with us for firewood that we would deliver at a later date. With these orders in hand we prevailed upon the parents of a friend to loan us the money necessary to buy two truckloads of firewood in northern New England. The one thousand dollars that we were thoughtfully loaned seemed of epic proportions at the time; I remember to this day the anxiety that came from borrowing it. Without the support of my new business partner, I doubt that I would have had the courage needed to even seek the money we needed. Finally, after much toing and froing, we succeeded in maneuvering two fully laden eighteen-wheeled tractor trailer trucks into Harvard Square and unloaded thirty-six cords of wood just outside the school president's office. We sold it all in the space of but a few hours, recouping the loan in its entirety and earning as much money as a year's worth of bathroom scrubbing could provide. We had uncovered an entirely different way of earning our keep. Hooked on the adrenaline and the freedom that we had discovered, we resolved to find a new opportunity that offered considerably more promise. After much thought, we came upon an idea.

My partner had spied an advertisement in a local subway offering college students an opportunity to act as salesmen for a revolutionary product. Following up on the offer we soon found ourselves in the presence of the inventor, a smooth-talking former chemical salesman who had designed a fuel saving device for automobiles. He proposed to hire us as salesmen and integrate us into his one-man operation. We merely had to deposit the savings that remained from our firewood venture and buy a small inventory of his systems. We naively and enthusiastically accepted. The future was ours.

Within a few months, however, it became clear that while the chemistry of his approach was conceptually sound and endorsed by many (including my father), the system offered only erratically positive results. Its flaw seemed to lie in the design limitations imposed by the target market. As a result of this experience, we thought of redesigning the system so that it would reduce fuel consumption for small industrial and commercial boilers, a market with constraints that seemed better suited to the process's strengths. The inventor agreed to help us design a product to our specifications and granted us the exclusive rights to such a system in return for a small ownership stake in the business and an on-going royalty. With an investment of ten thousand dollars from a friend's girlfriend's father, we capitalized the small company that was to market the fuel saver and hold the rights to the idea that so captivated us. I was eighteen years old, and, boy, was I excited.

Building Fuel Tech was a challenging seven-year struggle that more resembled the lurching gait of a man who has had too much drink than it did any thoughtful plan. Two steps forward, a step and a half backward, filled with exhilaration, a sense of wild enthusiasm, and a heady dose of the fear of failure, I continued. Virtually everything changed about the company save its name. From the three of us who had begun together, only two remained six months later, and within a year and a half only I was still working full time at the company. The system with which we had begun had limitations of its own that required its complete overhaul, and the money that had sustained us during the early months quickly proved insufficient for our needs. By 1982 the broadly accepted projections of a tremendous rise in the price of fuel oil that were so common as our company began in 1980 were proving to be nothing more than fanciful thoughts. The company underwent one transformation after another as it struggled to surmount new obstacles, each requiring a new set of resources, a new level of energy, and a new level of insight from me. Through all the delight and disappointment, I worked on building the company even as I tried to earn my college degree. Energy, persistence, and boundless optimism seemed to be the fuel with which one powered a small business like mine.

As the fervor for reducing fuel consumption subsided, it became apparent that there was a tremendous opportunity for developing systems that reduced the air pollution caused by combustion. Reassessing our objectives, we sought to convert our firm into a developer of cutting-edge air pollution control technology. With great difficulty, we gathered new supporters to our reoriented program. Those who joined our firm as employees risked both their careers and their families' fortunes in the hope of finding personal fulfillment and financial freedom through our company's success. Investors who recognized the potential to further their social or financial objectives offered us support, and sometimes even provided it. Talented and experienced men and women from business, government, and universities in both the United States and Europe joined our firm as board members and technology collaborators. From painful mistakes and the wisdom of others, I learned, and the firm grew. Slowly I came to understand both the natural disorder of a fledgling enterprise and the critical role that a talented group of committed people can play in enlivening and directing it. I discovered the benefits of patient and understanding investors, and the pains of short-term and unprincipled ones. The dance of legislation, so critical for creating a market for pollution control systems, became comprehensible to me, if never easily predictable. The absolute requirement that a great company have a clear and principled culture with a steadfast dedication to excellence as a standard became obvious. I discovered the strengths of a legal system that respects intellectual property and contracts and the trials of being an employer subject without cause to irresponsible legal assault. Most importantly, however, I learned of the vast technical, educational, and commercial resources of America. Traveling broadly throughout the world and creating subsidiaries in a number of foreign lands, I always returned to marvel at my own country.

By 1987, when I sold my stock in the company and resigned as Chairman and Chief Executive Officer, Fuel Tech had developed a wide range of air pollution control technologies for use in countries throughout the world, received more than thirty patents, and built a staff that numbered some two hundred in North America and in Europe. Most of those who had been investors or employees in the early days had become millionaires, and the company was valued at roughly $200 million. I was twenty-five years old and had achieved my goals of being an inventor and building a technology company. I was financially independent. I had learned from some gifted people and had a great deal of fun.

Not surprisingly, there were also pains in the process of building Fuel Tech. In my inexperience, I hired employees and attracted investors whose objectives were at odds with those of the company.

Parting with even the most trying of people was intensely painful at times. The constant demand of attracting the financial resources necessary to support the capital-intensive technology development in which we were engaged was wearing. Responsibility for performance that all too often seemed out of my control was a new position for me, and at first I found it a bewildering change from the pleasant predictability of school. The drain of baseless litigation was often enervating and seemed a confusing abuse of the principles of due process that I understood only in their most desiccated form. Watching people change positions of principle so as to increase their short-term gain was eye opening and sometimes quite upsetting. Most difficult of all, however, was the all-consuming emotional commitment that my work had demanded, a commitment that sometimes left me too drained to be a good friend, a good son, a good citizen. The exhilaration of living my dream carried with it the sense at times that I was doing a high-wire act witnessed by my family, my friends, my customers, my employees, and, in a small way, by society and the ever-present lawyers. The seemingly endless mistakes that I made were each scarring because in my immaturity I did not expect to err so very often. I had enjoyed the process deeply but as it became clear that the technology that we had developed could be marketed more effectively by a firm with more resources than ours, I was happy and somewhat relieved to bring my time with the firm to an end. Although I take great pride in knowing that there are more than 160 plants operating worldwide with the systems we developed, reducing pernicious forms of air pollution, I must admit that I have never had an urge to go back. Reflecting on all that I had learned, I resolved to begin afresh and to build a world-class firm dedicated to leveraging technological innovation to the natural world's advantage.

Pondering the lessons of my first seven years in business helped me to establish a core set of principles that I would rely on and use to build a new company. Largely drawn from previous mistakes, they served me well in evaluating new opportunities. First, I was determined only to work on a subject that was in simple harmony with my own values. However necessary, money-making as a goal is neither uplifting nor truly sustaining. Money can be a means to an end; I wanted to focus on the end. Second, I was committed to doing what the Patent Office calls "novel, unique, and non-obvious work." My fascination with the promise of technology remained. Third, I was committed to having partners, true partners, in any future enterprise. The pace, complexity, and emotional pressure of a technology start-up demands the support and skills of a team. Fourth, I was committed to creating the world's finest firm in its field. A dedication to true excellence by the highest international standards requires resources from around the world, and the internationalization of technology and commerce offers this as a tremendous quality advantage. The risks that I would take would not be risks of quality. Fifth, I would remember the fantastic benefits that mentors offer. Wisdom, perspective, and grace can be great reservoirs from which to draw during times of trial—and I knew that there would be times of trial. Sixth, I would not let my work impinge on my ability to be a loving brother, son, husband, or father. Seventh, I would remember the energy offered by love—love of my work, the people I work with, and the mission itself. Eighth, I would try to take my work, not myself, seriously. I knew that I would make a great many mistakes but that I would have to admit them and move on if I was to achieve my goal. Finally, I would remember Goethe's maxim, "Whatever you dream, begin it, for boldness has genius, power, and magic in it." Luck plays a great role in achievement, but preparation and the magic of boldness can help.

The concept for the most valuable product that we developed at Fuel Tech sprang from work done at a utility-funded, nonprofit research center, the Electric Power Research Institute (EPRI). By licensing from EPRI, we were using one of America's most extraordinary resources—the wealth of technology available for license from government, university, and nonprofit laboratories throughout our country is unique and staggering. As I searched for the basis on which to build a firm focused on using science to revolutionize the nascent environmental industry, I began by using an informal network to survey the results of this monumental investment. Within two years, I was a cofounder of two firms, Energy BioSystems (EBC) and Molten Metal Technology (MMT), each of which had the potential to meet my objectives.

The mission of Energy BioSystems is to use the revolutionary new tools that the biotechnology industry has created to alter some of the fundamental dynamics of the energy industry. The processing of hydrocarbons to produce either fuels or petrochemicals, and the subsequent use of these products, would ideally rely on feedstocks comprised entirely of carbon and hydrogen atoms. Such materials would not only create high-value products and be relatively easy to handle, but their conversion would also result in dramatically lower pollution than that generated from hydrocarbons contaminated with sulfur, nitrogen, or vanadium. However, far from the ideal pairing, the hydrocarbons that make up the world's present feedslate are laden with a witch's brew of elements that make the refining and combustion of fossil fuels not only expensive but a horrific source of pollution. For example, it is the sulfur contamination in oil and coal, which the heat of combustion converts into a precursor for sulfuric acid, that is the primary source of acid rain, a painful form of air pollution that all industrial and industrializing economies wrestle with. The United States alone emits some fifty million tons of acid rain precursors each year. While Energy BioSystem's long-range program calls for it to introduce biorefining and biotreatment as a method to remove and recover all the trace materials from fossil fuels, the company's initial objective is to commercialize a microbial system for removing the sulfur. This system will be the largest-scale application of biotechnology to the energy industry to date, with the potential to upgrade crude oil and other hydrocarbons worth tens of billions of dollars each year.

Still in the development stage, Energy BioSystems is a small reflection of a national trend. With employment growth in America coming largely from firms with fewer than five hundred employees, the fact that these small firms can double in size each year, as EBC has done, is an important part of our society's ability to continuously expand our employment base. Moreover, since the preliminary technical work that uncovered a biological package with the potential to meet our requirements was done under a Department of Energy funded program and carried out at an industry-funded nonprofit laboratory, the Institute of Gas Technology, I believe that EBC's achievements highlight how the public sector and the private sector can complement one another in ways largely unavailable to countries that lack such a mighty investment in nonprofit research. In addition, the willingness and ability of the capital markets in the United States to fund a high risk/high reward attempt to create such a large-scale application of biotechnology through a series of investment rounds that have incorporated private individuals, venture capitalists, strategic corporate partners, and the public market is something that sets America apart from other nations. I believe that investments of social and financial capital in potentially explosive technology firms will pay rich dividends for America in the future, as they have in the past. While not all will succeed, each will itself catalyze new thinking. Just as importantly, no other country has the infrastructure and the attitude to create the range and number of technology companies that can be found in the United States. Entrepreneurship as an exhilarating and inspiring act of high-wire performance art finds a great source of material amidst America's technology leaders. Firms like Apple, MCI, Microsoft, Genentech, and Sun now join Hewlett-Packard, Intel, Motorola, Merck, and Disney as icons of American culture as well as tremendous sources of national wealth. Young people like myself all over our country aspire to the standards set by the leaders of these firms and their peers.

Despite the promise of Energy BioSystems, I have chosen to focus my energies on building Molten Metal Technology. Begun roughly five years ago from a technological base licensed through the Massachusetts Institute of Technology (MIT), Molten Metal is dedicated to the development and commercialization of a proprietary system for converting the eighty billion tons of municipal, industrial, hazardous, and nuclear wastes generated each year by the global community into industrial grade products. By uniting research laboratories and corporate sponsors the world over in a development effort that has committed almost $250 million, Molten Metal's Elemental Recycling System offers the potential to harmonize the objectives of industrialists and environmentalists by forcing a paradigm shift in the way society views waste. Still early in its corporate life cycle, I believe that the small size and relative frailty of MMT highlights well a set of issues that offers perspective on America's economic future.

Begun as an act of partnership drawing on the different skills and experiences of a series of talented folks, it was quickly clear that MMT's resources would not be sufficient for our aspirations for quite some time. We were determined to achieve excellence by focusing intensely on developing and commercializing the linchpins for our technology vision, seeking talented help on any subject that was less critical, and sharing any success generously with all those willing to aid in our mission. Like many before us, we began by leveraging society's rich support of our university system, for not only was MMT's genesis sparked by MIT but we have continued to draw heavily upon MIT's vast resources, seeking counsel from its professors, insight from its research, and employees from among its graduates. Indeed, the results of this relationship have caused us to initiate development programs with similar objectives at five other campuses across the United States as well. In addition, we operate training programs for our employees in conjunction with universities close to our in-house research facility and gain considerable advantage from the technical libraries and data bases of universities around the country. Our use of the wealth of America's public science goes far beyond this, however. We also frequently take advantage of the abundance of resources available from the more than 750 federally funded national laboratories. Whether using supercomputer time available through the National Institute of Standards or material handling skills that have been developed by Sandia, a Department of Energy facility, we consult daily with members of our government's research establishment. The costs of such collaboration are low, the support heartwarming, and the benefits immeasurable as we draw upon resources that no other nation can match.

In our search for thoughtful, constructive support, we have not only looked to nonprofit research centers. A web of strategic alliances with a number of America's and Europe's leading firms has given us access to assets and insights of important customers and vendors willing to serve as partners in the commercialization of our technology. Pledging human, technical, and financial resources, these companies have strengthened us with their questions, observations, creativity, experience, and confidence. While not every program has met with either parties' expectations, even difficulty and disharmony have been sources of learning. With the pace of technical development accelerating even as the costs increase, product development cycles must be short, and the relationships between firms large and small, technology developing and technology consuming, domestic and foreign, must and will increase. The decentralized nature of our economy enables those who understand this to rush to reap the benefits of partnering. For Molten Metal, the benefits are clear: its web of relationships not only gives us the effective resources of a firm several times our size, but more importantly, it lets us focus our internal development program on the issues that are absolutely critical to achieving our goals.

Integral to accomplishing the technology revolution that we seek is the recognition that both our competition and our markets are international in scope. To build our understanding of international markets and support our objective of pursuing the highest quality research available the world over, we operate development programs throughout the world. Overseen by an international board of technical advisors, this work provides us with opportunities to hire the most talented scientists in our field in addition to providing us with tremendously valuable technical and market research. As a result, although a small firm, Molten Metal presently employs people from more than twenty countries whose technical skills and perspectives are absolutely critical to driving the pace of our work. These people join Molten Metal because of the challenges and opportunities offered by our work. They are willing to leave their own native lands, however, in part for the promise and welcome that they expect from America. Our ability to consistently attract gifted people willing to risk painful cultural separation in order to take their chances amidst the chaos of capitalism has been a tremendous source of strength for our firm. Likewise, our nation's ability to attract talented and ambitious people from foreign shores to our own is a tremendous source of strength for the United States.

Drawing upon resources from partners in both the public and private sectors is necessary but not sufficient to foster the explosion in high-tech firms that can play an important role in leading America through the third industrial revolution. The capacity of our nation and its financial system to provide the capital these firms need for development and rapid growth is also critical. For companies like Molten Metal, already among the most richly financed environmental technology start-up firms ever, the fundamental strength of America's financial markets will continue to play a crucial role in our development and growth. Without our nation's vast reserves of patient, thoughtful, creative, and aggressive equity, industries at the leading edge of technology in fields such as overnight shipping, supercomputing, energy storage, biotechnology, and pollution control would never have gotten off the ground. America's ability to allocate capital quickly and wisely in these nascent fields is a tribute to our financial community and a tremendous source of national advantage. No other country can match the dynamism of our capital markets, particularly in their ability to finance rapidly growing firms. Unless we damage this extraordinary resource through indolence or unwise legislation, this should long remain an almost unique tool of American capitalism. For MMT this has certainly proven an impressive source of strength.

Powerful and supportive partnerships, extraordinary technology, dedicated customer orientation, intelligent financing, and even talented employees are not enough to build a successful company. To be great, a company must build a culture on an inspiring set of values; it must understand its mission and develop and follow a clear strategy for achieving it. It requires leadership with a well-conceived and communicated vision and the will, courage, persistence, creativity, and humor to overcome the inevitable and painful obstacles that will challenge its realization. Great companies require great culture, and a cornerstone role of the entrepreneur is to establish a goal and to construct and fiercely protect a culture that unites, directs, and inspires those who are focused on reaching it. It follows that if the dynamism of America's economy is drawn from millions of committed, talented citizens pursuing their interpretation of the American dream, then a critical role of our government and our public leaders is to nurture a culture that educates, unites, directs, supports, and inspires these citizens to achieve this objective, in part by reminding them that the American dream is not something that one wins but something that must be earned. Equally, we must all be aware that fortune plays a role in achievement and that honest, hard work that leads to failure is no disgrace because merit is in the striving. Our national willingness to tolerate failure has been a reservoir of deep advantage plumbed by such American role models as Presidents Lincoln and Truman, Henry Ford and Remington, but it seems to be on the wane. Mistakes, foibles, or pure failure are now no longer only feed for media but also provide a steady diet for litigation-happy lawyers. As merely posturing to avoid failure can diminish the magic of boldness and dampen the raw potential of idiosyncratic creativity, the leaders of our country must reverse trends that increase the pain of honest failure. If continuing to be the world's economic leader is to remain a national goal, the risks that attend those who strive mightily must be risks that we are prepared to accept.

Indeed, our leaders' recognition of and dedicated support for conditions that honor the individualism, imagination, and initiative of all Americans will be critical in determining the future of our society. America's advantages as we enter the twenty-first century are simply fantastic. In fields as diverse as mining and energy, forestry and agriculture, America has a natural resource base that is impressive by any international standard. It has tremendously effective transportation and communication systems and a university system that is the envy of the world. Drawn in large part from a base of government-funded research, America's technological resources are simply stunning. From aerospace to telecommunications, from pharmaceuticals to software, from chemicals to microprocessors, the nation's riches in the vanguard of the economy of tomorrow are almost immeasurable. Fueling the world's largest and most productive economy, the capital system of America—its fluidity, its vigor, its intelligence, and its sheer power—has become itself a formidable competitive weapon. Even the popular culture of our land, from television programming and motion pictures to rock videos and fast food restaurants, is a source of impressive wealth and influence. With a political system that is both inclusive and stable, a work force that is educated and motivated, and a social system that is dynamic and diverse, our country has the capacity to deal wisely and effectively with the problems that confront it. The United States' present position is one of unparalleled opportunity and is a direct result of Americans' passionate commitment to a set of ideological standards that value freedom, individualism, and creativity as national treasures. Our dedication to a social system that rewards excellence and provides each citizen an opportunity to demonstrate it may be the most important characteristic in determining our country's economic future. Supported by the love of our families and the willingness of our communities to sacrifice for our education, giving thanks to the countless men and women who built America into the land of opportunity while wisely seeing that this ideology animates their construction, millions of young people like myself are offered a future that holds the promise of dreams that have come true. It is exciting to contemplate.

Thomas D. Cabot

A Short History of Cabot Corporation

T HE CABOT FAMILY GOT THEIR START in the carbon black business in 1882 when Sam and Godfrey Cabot (my father) purchased a small works in Worthington, Pennsylvania. Their father, Dr. Samuel Cabot, was a prominent surgeon who had graduated from Harvard Medical School in 1839. He was also an amateur ornithologist, whose large collection of bird skins led to the creation of what is now the Museum of Science in Boston, Massachusetts.

The older son, Sam, graduated from the Massachusetts Institute of Technology in 1872 and then studied in Europe at the Zurich Polytechnicum. He visited various chemical works in Switzerland and Germany and became interested in products that could be profitably made from coal tar. On his return to the United States, he bought a large vacant lot on the waterfront in Chelsea, Massachusetts and set up a laboratory. Godfrey graduated from Harvard College in 1882 and joined Sam in his venture in addition to doing some work as a consulting chemist and analyst.

Sam contracted to buy coal tar from the gas works in Boston. The distillate from the tar was used in such products as household disinfectant, sheep-dip, wood preserver, and shingle stain. The heavy pitch, which was the residue from the distillation, was used to make a fine powder known as lampblack. This powder was produced in a low, long, brick shed with a metal roof; there was an adjustable door at one end and a high brick chimney at the other. An iron pot at the door-end of the shed contained the pitch, which was set afire and then allowed to burn for several days. When the

Thomas D. Cabot was Honorary Chairman of the Board of the Cabot Corporation.

135

materials were exhausted, the works were left to cool for two days, then the soot was scraped from the walls and ceiling and loaded into jute bags; in all it was a very messy process.

One day a salesman brought Sam and Godfrey samples of carbon black that had been made from the gas fields in Worthington, Pennsylvania. At this time, the major use for carbon black was as a pigment in inks and paints. Sam visited these works himself. After testing the samples that he collected, he quickly recognized that the samples were much finer than the lampblack that he and his brother were producing and would be far more suitable for the ink used with modern, high-speed, rotary presses. Sam wanted to be able to sell this better quality of black and so he sent Godfrey to negotiate for the purchase of the plant, which he did successfully for a few hundred dollars. That winter, Godfrey decided he wanted to take additional courses in chemistry in Europe, as Sam had done, and so he spent a year at Zurich Polytechnicum. When he returned to the United States, after two more years spent traveling through Europe with his cousin Harry, he found that the Worthington plant had been shut down. There was not enough gas, the raw material used to make carbon black, to keep the plant operational. Godfrey visited the plant and recommended that they move it to a newer gas field near Kittanning, where gas was more abundant. Sam was unwilling to invest further in the plant, so Godfrey bought his half interest and moved the plant to the new site.

The output at the new site increased and, although Sam bought some of it, Godfrey soon began to sell much of it himself. Since the business kept him away from home much of the time, he opened an office in Boston to handle his business affairs while he was away. He never had any formal training in business and knew nothing of accounting and double-entry bookkeeping. His books consisted of only a checkbook or blotter in which the checks were listed. The business was a sole proprietorship and household accounts were included in the same books, a distinction being made by a capital letter on the stub of the check as to whether the transactions involved the plant (W) or household (H).

Godfrey had a strong sense of duty and allowed himself little leisure; he worked very long hours and expected his employees to do the same. He treated all his employees as hired hands who were not allowed to make decisions but rather were instructed in detail as to everything he wanted them to do. He paid low wages and hired employees who had very little formal education. In fact, with two exceptions, none of his employees had gone beyond the eighth grade in school.

By the mid-1880s, the carbon black business had become very competitive. Godfrey was approached many times to join other manufacturers in price agreements or in agreements to curtail production, but he did not trust his competitors. In 1893, when he decided to take some graduate courses at Harvard, he chose to sell his black through the New York firm of Binney & Smith. However, that year was a difficult one economically and he believed that Binney & Smith were giving preference to the output of other firms, ones for which they were agents and in which they also had financial interests. He canceled his sales arrangement with them, dropped his Harvard courses, and spent much of his time on the road selling his own output.

In addition to his sedulous sales efforts, he returned often to the works in Pennsylvania and busied himself securing his raw material source by making new gas contracts and acquiring wells, gas lands, and pipelines. He also began to sell gas to heat houses and stores along his lines; he gave the gas away free to farmers and to other land owners in exchange for rights of way for his new lines. Although some believe that he was not a good businessman, he was certainly a good negotiator and shrewd buyer.

After the election of President McKinley in 1896, the business was prospering, and Godfrey decided to go to Europe, where he appointed new foreign agents and visited the gas fields of Russia. At that time, he had several small works in northwest Pennsylvania. He knew that cheaper gas was available in West Virginia, and so he decided to build a much larger plant in Grantsville on the Little Kanawha River.

The Grantsville plant, built in 1900, was the largest in the world. It consisted of a hundred tent-shaped metal sheds; these were circular, with a strong metal pole at the center, and twenty-four feet in diameter. Hanging by diagonal rods from the top of the poles were the heavy cast iron plates on which the black gathered. The scrapers, the hoppers, and the piping to supply the gas were all rotated by a central bracket and chain. It was a poorly designed apparatus and needed frequent adjustment and repairs. Nevertheless, it worked well enough. Godfrey was able to buy the gas at the well for no more than 1.5 cents per thousand cubic feet; he put in his own network of pipelines to carry the gas to the central works. It was this network of pipelines that made him rich. It gave him a virtual monopoly on the gas produced in much of the valley.

Grantsville had one great disadvantage, however. The mud roads leading down the valley to the railway were impassable for heavy wagons during much of the year. The black, which was mostly sold by the barrel in those days, had to be taken down the river while it was flooded. Through more than half of the year, there was not enough water to float the shallow river boats, so Godfrey built warehouses downriver at Creston and filled these when the river was navigable. The oak barrels for the black were built in a cooper shop next to the plant, and if the barrels were not ready, production had to be curtailed. Filling them was a very messy business: the black was simply shoveled into the barrels. Later, the black was stored in paper sacks, which were lightly pressed and packed into wooden boxes.

The first well Godfrey drilled for himself in West Virginia was the Mathews No. I, near Grantsville, which struck gas in 1899 and still continued to produce gas in 1970. After finishing construction on his large carbon black plant in Grantsville, Godfrey bought a smaller one across the river. He later moved one of his Pennsylvania plants, which had been shut down for lack of gas, to a site at Nancys Run in Roane County, near Spencer and west of Grantsville. Spencer was the county seat and a city of about one thousand inhabitants. It was the railhead of a branch of the B&O, connecting with its main line at Ravenswood, near Parkersberg; Nancys Run was the last station on the railway before Spencer.

Using larger pipes than he had ever used before, he connected his Grantsville network of pipes to supply both the works at Nancys Run and the city of Spencer. It picked up gas from many wells along its route and was extended westward with smaller pipes to pick up gas in the western parts of Roane County. Just before the start of World War I, he began construction on another small plant near Shreveport, Louisiana, but the plant was never completed.

* * *

When the Wright brothers first flew at Kitty Hawk in 1903, Godfrey was very excited about their invention. He had inherited an interest in flight from his ornithologist father and had followed closely the glider experiments of Otto Lilienthal in Germany. When the Wrights' flight was announced, he wrote to the Wright Cycle Company and offered to help finance further experiments. He also wrote to President Theodore Roosevelt, a cousin by marriage, and to the two senators from Massachusetts, urging them to pass legislation that would allow the United States Army to buy the rights to the Wrights' invention. He was fully convinced, as he said in these letters, that man would ultimately fly faster and farther than the birds and that mechanical flight would be the most important weapon in future wars.

In 1914, when war broke out in Europe, he was so certain that airplanes would play an important role that he bought a plane of his own, designed and built by Starling Burgess at a yacht yard in Marblehead. The design was similar to a British plane made by Dunn and was given the name Burgess-Dunn. It had an extreme sweepback, a highly cambered wing, no tail or rudder, and only two ailerons, each activated by a stick in the cockpit.

I studied aerodynamics at the Massachusetts Institute of Technology in the spring of 1915 under an Army engineer named Alexander Klemin. I believe it was the first time that a course in that subject had ever been given. It was mostly concerned with cambered airfoils with a high lift-to-drag ratio and the instability of such airfoils caused by the center of lift moving forward as the angle of incidence decreased. When the Burgess-Dunn plane my father ordered was delivered in the summer of 1915, I pointed out to him that it would probably plunge into the sea if it were glided at too low an angle and that he should not maneuver it too sharply.

When I returned from military training that summer, I began flying lessons and learned to repair the engine of the Burgess-Dunn plane, which was in constant need of maintenance. When the United States entered World War I, my older brother Jim and I trained at the Curtiss Flying School in Buffalo to be military aviators and officers in the Army Signal Corps.

Godfrey had learned to fly his Burgess-Dunn plane in the summer of 1915. It took off and landed on floats, and he had a hangar on Misery Island, in Salem Harbor, with a marine railway that he used to launch and haul out the plane. He persuaded the Massachusetts legislature to establish a Naval Air Militia and received a reserve commission in the US Navy. In April 1917, when President Wilson declared war on the Kaiser, Godfrey was a full-time Naval officer and forbidden by law to conduct any business. The Navy put him in charge of a flying school in Marblehead; two retired chief petty officers helped him to train men from New England colleges to become naval aviators. The Marblehead school was abandoned in the fall and Godfrey was sent to a larger school at the US Naval Air Base near Norfolk, Virginia, where he helped Rear Admiral Fiske with the design of a torpedo plane.

After World War I, Godfrey built a carbon black plant in Cedar Grove, West Virginia, on the Great Kanawha River, east of Charleston and near the main line of the C&O. Steel channels were used to collect the black instead of the cast iron plates that were used at Grantsville. He then built a smaller plant in Boon County, south of Charleston, but his interest in the business soon waned. A substantial portion of his carbon black business had been shut down during the war years and production in the East was destined never to regain its pre-World War I position.

He moved to Washington, where he founded and became the first president of the National Aeronautic Association. My brother Jim went to Charleston, West Virginia, and assumed control of Godfrey's properties in that state. I completed my training as an engineer at Harvard and MIT and moved to Spencer, West Virginia, to handle the rebuilding of the pipeline.

Jim received dreadful, long letters from our father, telling him in great detail everything he must do in Charleston to save the business. I realized the futility of answering his letters and decided to follow my own lead.

In the summer of 1921, I returned to Boston. It was there that I learned that Godfrey's bank accounts had been expropriated by the Internal Revenue Service and that he had hired Merrill Griswold, the son of a Cambridge neighbor who had graduated from Harvard Law School and joined the firm of Gaston and Snow, to handle the situation.

Merrill confided in me that he was frustrated with Godfrey's obstinacy and that unless I could do something to help, US marshals would take my father to jail. He suggested that I hire a good accounting firm like Lybrands and an appraiser like American Appraisal Company to estimate the value of the properties as of March 1, 1913, which was when the income tax became law, and to prepare financial statements for the intervening eight years. I knew that it was useless to consult my father, and since I had power of attorney, I decided to go ahead and follow Merrill's advice.

When the tax bills arrived, Godfrey was irate, but he agreed to Merrill's recommendation that we incorporate the sole proprietorship. On October 1, 1922, all of Godfrey's carbon black, oil, and gas business properties were transferred to the Company in exchange for all of its stock. Thereafter the business was known as Godfrey L. Cabot, Inc., a Massachusetts corporation; Godfrey became the president, I the treasurer, and Jim the vice president. When Jim died in 1930, our brother-in-law, Ralph Bradley, succeeded him.

Before Godfrey moved to Washington, a Cambridge neighbor consulted him as to the future of the neighbor's daughter. She had fallen in love with a young man named Winslow Duerr, who was a teacher and earned a salary of only $100 a month. The neighbor wanted to know whether Godfrey thought the daughter should be allowed to marry him. When Godfrey learned that Duerr had graduated from Harvard *summa cum laude* and that his concentration had been in chemistry, he immediately said that he would offer him a job for $200 a month. When Duerr accepted, I was told to put him to work.

At about that time, Godfrey had been persuaded by a business promoter named Howell to invest in a natural gasoline plant in Salem, West Virginia, that would extract natural gasoline from casinghead gas being used in several glass factories near that town. Godfrey bought the major part of the stock and appointed me treasurer, releasing Howell from that position. I in turn appointed Duerr superintendent and let him run the company, but Standard Oil, who owned most of the gas passing through the plant and had agreed to take the product, found it too volatile to put into its ESSO brand motor fuel. We lost our market and the plant proved to be unprofitable; by 1925 it was shut down.

During World War I, carbon black sold for as much as thirty cents a pound. After the postwar boom, so many small newcomers had entered the industry that the price dropped to just five cents, forcing many of the newcomers into financial difficulty. We helped to finance some of them and became sales agents for others. In 1924, most of the independent companies had decided to merge and formed the United Carbon Company. Binney & Smith became the sales agent for more than half of the output of that company.

In August of the following year, we were notified by the leaders of the United Carbon Company and Binney & Smith that effective January 1, 1926, no carbon black would be sold for under eight cents per pound. Such a mandate was anathema to the Cabots, and we decided to go into the carbon black business ourselves in earnest. The decisions that were made at this time determined the course of the company's business for the next twenty-five years and resulted in the migration of the business from the eastern part of the United States to the Southwest.

Duerr and I went to Texas and over the course of three weeks tested the gas at more than twenty of the largest natural gasoline plants, where the gasoline was extracted from the gas and the residue was flared into the open air. The two most productive plants in Texas with a high quality of residue were the Humble plant at Breckenridge and the Phillips plant at Eliasville. We made contracts with both companies to buy their residue gas under the name of a subsidiary equally owned by us, which we would manage. The Breckenridge plant, where we first established our southwestern division headquarters, was built largely with materials salvaged from the Shreveport plant that Godfrey had partially completed before the war. Because the gas supply was somewhat marginal in quantity, we required Humble to guarantee that this supply would remain adequate for our needs. The wells they had expected to drill were disappointing, and after the plant had run for less than a year, we were obliged to ask Humble to make good on their guarantee. They bought our half interest and we shut the plant down. In contrast, the Texas Elf Carbon Company, which was created at the Eliasville plant jointly with Phillips, proved very profitable and gave us a 100 percent return on our investment in the first year. The Eliasville plant quickly doubled in size, and Texas Elf Carbon Company started to build another plant south of Pampa, in the Texas Panhandle, which became the largest carbon black plant in the world and our new southwestern division headquarters.

Before the second part of Texas Elf Carbon Company had been started, however, a subsidiary of Cabot built a very large carbon black plant at Skellytown, west of Pampa, which was also very profitable. Between 1925 and the onset of the Great Depression, we reestablished ourselves as a major producer of carbon black by building nine plants in Texas and Oklahoma. Most of the plants thrived, and the percentage of the world industry that we controlled had risen from less than 1 percent, at the end of the war, to about 15 percent, although we had to borrow heavily to do this. All of these plants were channel construction, and, except for the first two, they were built from a new design that used our engineering training. Our capital costs were reduced to less than 60 percent of the cost of our competitors' plants and to about 60 percent of our overall operating costs.

One improvement we made was to use electric welding in construction, which was ideal in building our low sheds and tables. This type of welding had not been approved by the engineering societies for the construction of buildings and bridges. But if a weld broke, a strong man could hold the structure in place for as long as it took to reweld it, and it was much cheaper than the bolts and rivets used to connect the steel parts in our competitors' plants. We also introduced unit electric motors instead of the central gas engines used by our competitors. The central gas engines limited the size of a unit to about thirty buildings and were driven from a central power source by shafts, chains, and sprockets; if anything broke, the entire unit would be shut down and repairs would begin only after the unit had cooled, approximately twenty-four hours later. It would be at least forty-eight hours before the unit was functioning and able to make uniform black again. Another major saving that we enjoyed was that our plants were scientifically designed to have a uniform pressure at each of the several million small gas flames. This was accomplished by equalizing the pressure drop using small pipes instead of the huge pipes that our competitors often used. Also, our style of construction allowed us to design buildings and units that were ten times larger than our older models. We were able to reduce our labor costs in the packing rooms to about half the cost of the smaller units in our older designs. The reputation for advanced processes and technology established by our company during these years remains one of our greatest assets today.

Originally hired as a research chemist, Ned Billings was responsible for our entire carbon black business in 1930 when both Duerr and my brother Jim died and when I was diagnosed with general sepsis and told that I would never be able to go back to work again. In the summer of 1932 Billings moved to Pampa, where he discovered a way of churning carbon black when it came out from the hot houses so that a portion of it would become partly pelletized; the pellets were then removed, and any black that had not been pelletized was recycled until all of the black was in pellet form. Trade-named the SPHERON process, the pellets were dustless, about one to two millimeters in diameter, and two to three times denser than the fluffy black that came out of the hot houses. The bulk density was about equal to the heavily compressed carbon black that was then being supplied to the rubber industry. We tried shipping the pellets in bulk, using an ordinary cylindrical tank car with an Archimedes' screw in the bottom. The black would collect at the center of the car, where a valve would drop it through a sleeve connected to a hopper opening between the tracks, then to a conveyer, and finally into a large tank within the factory.

We promised our customers that we would price the bulk black at a quarter cent per pound less than the black sold in bags, passing our savings along to them. Two of the larger tire companies decided to make the initial investment necessary to use this bulk black in their plants; they were thoroughly satisfied with its advantages, not the least of which was that it was cleaner to use. The other tire makers, having been promised by Binney & Smith that their firm would match our prices, decided not to install bulk handling equipment. A price war ensued. Every time we offered bulk prices at a quarter cent discount, Binney & Smith matched our price. When the price dropped to the point where we were not able to cover the incremental cash price of operating a plant, we were forced to notify our customers that we would no longer be able to make good on our promise of a bulk discount. We promised to discount the carbon black again once the price returned to normal. The price, however, did not return to normal until after World War II; it was kept artificially low first by the National Recovery Administration and later by the Office of Price Administration.

In the late 1920s and early 1930s the Great Depression sharply reduced carbon black volumes and prices, slowing the growth of our company. By 1939, however, our sales were nearly $7 million, with carbon black accounting for a little more than one-third of the total.

The stirring process that we used to create pellets was soon outmoded by our competitors, who had found a much better and cheaper way to make pellets: they wet the black with a very dilute solution of corn starch and then dried it in large rotary dryers, which made firmer and denser pellets. We soon adopted this method at all of our plants. This process could not be patented, however, because it had been used previously on other materials. A patent for making dustless carbon black by stirring it into a mixture of water and gasoline was found in the Binney & Smith archives. After the mixture settled, the carbon black was attracted by the bubbles of the gasoline, and, when separated from the water, rose to the surface. The water was then drawn out from below and the gasoline was boiled off to be reused later. However, the process was far too expensive ever to be used commercially.

Our patent was for a dustless product and we earned a small amount of income because of this claim. One of our licensees, however, was challenged for infringing on the Binney & Smith patent that preceded ours. Our licensee was sued by Binney & Smith. We paid for his defense and won in the District Court but lost in the Circuit Court. We received *certiorari* to carry the case to the Supreme Court. We won and thought that the matter was settled. However, while we were occupied with the war effort, Reid L. Carr, a prominent New York attorney who had taken over the leadership of Binney & Smith, managed to alter slightly the wording of the Binney & Smith patent; he sued us claiming indemnity for all the pelletized carbon black that we had sold through the twelve or more intervening years. He won in the District Court, and we knew that if we appealed we would most likely lose in the Circuit Court. We knew we had little chance of getting *certiorari* to bring the case to the Supreme Court. The outlook was very bleak indeed.

We were financially extended because we had decided to keep our carbon black plants running during the war, even though the demand for the product was low. We stored the surplus carbon black in a row of warehouses nearly a mile long, knowing that the demand for it would increase once the Baruch plan for creating a large synthetic rubber industry materialized. This plan was devised in response to government-imposed restrictions on rubber consumption that resulted from a cutoff in natural rubber shipments from Malaysia and the East Indies. We had predicted that there would be a shortage of carbon black, essential for reinforcing both natural and synthetic rubber. Throughout the war we pleaded with the War Production Board (WPB) for permission to increase our production capacity to meet the anticipated demand for carbon black by the new synthetic rubber industry. But the government could not decide on what kind of carbon black would be suitable, and we did not receive permission to build until late in 1944. Unfortunately, once we received permission, construction of the plant was delayed because Charles Eliot, a childhood friend who had been appointed by President Roosevelt to head an office in Washington concerned with the postwar economy, stopped our steel shipment, claiming that gas would be needed after the war. I convinced him that the gas was being flared and would continue to be wasted in the open air until our plant was completed. We received the steel we needed and built the plant, which produced a black in gas furnaces that was much too coarse for the treads of tires but suitable for the sidewalls. While carbon black that was made through the channel process continued to be in demand, the newer furnace process produced a black with the properties needed for synthetic rubber in nonwear uses.

On one of my trips to Washington I visited with Bradley Dewey, who was in charge of the synthetic rubber industry that was expected to be in operation within a year. I told him that I felt that his rubber program was going to be crippled by a shortage of carbon black. He wrote to Donald Nelson, who was in charge of the WPB, and told him about our conversation. The letter was very persuasive; in no time at all, the carbon black industry received the WPB's full support. We were encouraged to employ Stone and Webster to build a plant at Guymon, Oklahoma, at the government's expense, that would produce channel black. The terms of our agreement allowed us to lease the plant and to run it at a profit of one dollar per year, no more, no less. We provided the organization necessary to build the plant, supervised its construction, and operated the plant for the government under this lease. In this plant, an oil feedstock, first used as a supplement to natural gas in the furnace process, became the primary raw material that was used. The transition to an oil furnace process launched a new era for Cabot as an international company.

I did not receive the lease document until nearly a year later. It had been approved by our counsel, but I noticed a clause that required us to indemnify the government against patent suit. Realizing that we most likely would lose the Binney & Smith patent suit brought by Carr and that this would bankrupt us, I refused to sign the lease. I soon learned that the terms of the lease were required by Congress and could not be changed, except by their act. Great pressure was put on us to comply, but I still refused to sign. A very distinguished patent attorney, who was serving as a dollar-a-year man working for the WPB in Washington, paid me a visit. He first told me that he did not blame me for not signing the lease, that if I were his client he would have advised me not to sign it. He then asked if I would settle the suit for $1,000; I said yes. Two days later he informed me that the suit was settled according to my terms; Carr had signed the document himself, which deprived him of all income from the industry. Carr was told by the president that he could not collect from the patent because the government would take in taxes double the amount he received. The president's war powers were almost unlimited. I signed the lease, and a few months later the synthetic rubber industry, planned by Baruch, got off to a flying start.

After the war, Carr tried to retaliate by buying control of a company that had been founded by Henry L. Doherty. The huge utility company, headed by Doherty before the war, had attempted to make a reinforcing carbon black from gas in a ceramic furnace but had succeeded in making only a much coarser semi-reinforcing black. Doherty obtained the patents on this process and built a plant near Pampa. Our shops in Pampa provided most of the steel work for the plant, and we came to know the manager, Wright, very well. Doherty's widow later acquired the company and wanted to sell; Wright did not want Carr to gain control of the company and so joined forces with us. We were certain that Carr wanted the company so that he could sue us for infringement, for using a somewhat similar design at our Ville Platte plant. We gained control of the Doherty company, avoided a potential suit, and hired Wright to work for us.

We survived World War II with some difficulty. With no rubber, sales of carbon black dropped substantially. Because we designed and built our own plants in the late 1920s, by 1940 we had as many as four thousand workers whom we had trained to shape, cut, and weld steel—the skills most needed for building ships. During the Depression we had kept many of these workers busy building pumping units and other equipment for the oil industry. Now these men were needed to build ships for the war, but our headquarters and shops were located too far inland to be of much use. We obtained some contracts to build landing craft, pontoons, and smaller items, and we talked with Army Ordnance about a contract to build big guns. At first this approach to Army Ordnance was fruitless, but in the fall of 1942 we were the first to begin production of the highest qualities of steel used in making big guns, though we did not make a profit until the war was over. When the Korean War began, we went back into this business and developed a scheme for making guns at about a 20 percent savings as compared with our competitors. We made a lot of money from the Korean War and from the subsequent rearming of Europe after the formation of NATO. By 1960, guns had been largely replaced by missiles, and we dropped out of the munitions business.

* * *

At the beginning of World War II, Phillips Petroleum introduced a brand of carbon black that was too coarse to be of interest for tire treads. It did have the quality, however, of being easily compounded into rubber that could be used to make strips for products, and they developed a small market for it.

At the end of the war, Phillips sent out samples of a better quality of black. When we received these through customers and had a chance to test the black, we found it was of a much finer quality and fully reinforcing. This was alarming because we knew that the carbon black business was really very inefficient—less than 3 percent of the carbon in the gas remained in the product—and all of our experience with furnace blacks showed that the black produced was too coarse to be really reinforcing. Even more troubling was the fact that black made from oil could be made very economically at any sea port, whereas channel black was made in North Louisiana and in the Panhandles of Texas and Oklahoma, where the gas was cheap. The gas could not be transported far by pipeline without running into heavy costs. I could see that the new furnace black industry had the potential to spread worldwide, leaving our channel black plants behind.

I immediately asked our R&D department, headquartered in Cambridge, to determine how such a good black could be made in a furnace. Our pilot plant crews in Texas were given the same challenge. When we discovered that the secret seemed to be in the aromaticity of the raw material and was not just a matter of furnace design, we began to build new plants. We knew, of course, that we would probably run into patent problems but decided we would have to take that chance and hope that we could negotiate later for a license at a reasonable fee.

In the late 1940s, with plants under way in the United States, we were ambitious to build abroad, and I went to England to see what was going on there. I discovered that the English had sent engineers into Germany at the beginning of the Occupation to learn how the Germans had produced some of the critical materials they needed to win the war, materials that they had previously imported. We found, to our surprise, that they had made a good grade of carbon black, almost as much as they needed for the war effort, by enriching artificial gas with a liquid made from solid anthracene residue, which was in abundant supply because it was a by-product of manufacturing U-boat fuel. The English had appointed a government commission to make use of this process in producing all the carbon black that was needed in England.

Although no patents had been issued, I felt confident that the process we were about to use would be much better for the English economy. It took about six weeks to convince the English scientists. When I returned home, I found it even more difficult to convince my associates. The consensus was that England would never recover from the war. After weeks of frustration, I announced that I would take a license from our company, use its technology, and build a plant myself in England to make this new carbon black. I would even move to England and give up my position in Boston. But Billings had recently died, and it was apparent that no one else could run the company. My son, Louis, went to England instead and built a plant in Stanlow, on the Mersey estuary. The plant was very successful and led to our starting similar plants throughout the world.

Our furnace black plants in the United States came on-line before Phillips's, and even though no patents had yet been issued, the patent office notified us that many of our claims were being claimed by others. Phillips approached us and wanted to make a deal, but their terms seemed exorbitant. It was apparent that there would be a major infringement suit; to avoid the high legal costs of such a suit, we finally came to an agreement that provided not only for cross-licensing but for pooling our patents with those of Phillips. Phillips would be the licensing agent and receive 80 percent of anything collected from third parties. Our research efforts in the carbon black field were to be merged for a minimum of three years. After the contracts were signed and the books examined, we realized that we should have made a much better deal. Our patent application was much stronger than theirs, and our patent position and knowledge of the process were much better. We were committed to three years but immediately notified Phillips that at the end of three years we would break away. Today, Phillips is only a small factor in the industry. We are the largest carbon black manufacturer in the world and almost twice as large as the next largest one, manufacturing about 30 percent of the world's production.

* * *

After the lend-lease program with Russia got under way, we learned that one of the most important elements missing from the war effort was vitamins, especially in Russia. A young lawyer, Titus, within our organization went to Washington with an idea of how we could help. He obtained permits that would allow us to extract vitamins from fresh alfalfa growing on the US side of the Rio Grande delta in Texas. We purchased six hundred acres of land with alfalfa already growing on it and put our own money into the plant. We soon learned that a process for making ascorbic acid from sugar had been invented and could be used to produce this vitamin at less than one-tenth of our cost. We changed gears and started to produce carotene B, which was then the principal vitamin in demand in Russia. We converted the plant to make this vitamin just as the war ended, but we soon found that our vitamin had a grassy taste, making it nonsaleable. The country next went through a craze for chlorophyll—based on popular whim, not proven science. We converted the plant to make chlorophyll, but when the craze subsided there was little market for this product and the plant was finally torn down.

We have had many other attempts at diversification since then. Some have been unsuccessful, while others have been very profitable. Our Pampa gun plant was converted to produce steel casings, mostly for the oil industry. Unfortunately, we soon found that our costs were high and the market very limited because of our distance from the coast or industrial centers. That plant was eventually shut down. In 1970, we got into the metal industry in Kokomo, Indiana, by buying from Union Carbide a large plant that made nonferrous metals, mostly cobalt alloys, which they were operating at a loss. We formed a new company, the Stellite Division, eliminated the unprofitable aspects of the plant, and ran the plant quite lucratively. Over the next fifteen years, additional specialty metals products and businesses were added to the Stellite Division nucleus. Indeed, it was the spring-board for the company's involvement in several related busi-nesses until the mid-1980s when the Division was sold.

Not long after the war we also became involved in making titania, a white pigment in large demand, using a process that did not seem to infringe on the profitable process invented by DuPont. We operated a pilot plant in the Alsace region of France for one year and then decided to build a plant in Ashtabula, Ohio. After the Ohio plant was fully operational in 1963, we identified some environmental problems and decided in 1972 to sell the plant to New Jersey Zinc.

During the 1960s, we also sold carbon black to a company that mixed it with polyethylene in a batch process. We acquired this plastics dispersion business, which had plants in both the United States and the United Kingdom, and distributed the product to electric telephone cable suppliers and to plastic pipe and pipe fitting businesses. Despite our efforts to develop experimental plastic poly-mers and compounds, the US plants were unsuccessful, and we sold them. But the subsidiary in England prospered, largely because it continued to operate using a batch process. That business, the precursor of today's Plastics Division, remains an integral part of our company.

One of the most successful diversification ventures from the 1950s started with a license from the Germans, who wanted access to some of our carbon black patents. In exchange they gave us a patent on an invention for making a fume of silica that was not very different from the carbon black process. Silicon tetrachloride was burned in a special furnace in an atmosphere of hot steam to produce smoke. The resulting residue, fumed silica, is a unique chemical with a wide range of uses, including thickening, dispers-ing, flow control, and reinforcement. Since opening our first plant in Tuscola, Illinois, in 1958, we have expanded our business glob-ally, with plants in Germany and Wales. Today, we rank number one in the United States and number two elsewhere in the manufac-ture of fumed silica.

Another profitable venture originated in our own laboratories: a patented ear plug. These hearing protection products are sold today in all parts of the world to a wide array of industrial, consumer, and specialty markets. This led us to produce other noise suppression and vibration suppression products. In 1990, we merged our E-A-R Division with the newly acquired AO Safety Division of American Optical Corporation to form the Cabot Safety Division, which provides other safety devices including face shields and safety eye wear.

One of our early metals ventures is still growing today: the production of tantalum powders and alloys of other heavy metals. Known as the Cabot Performance Materials Division, this Pennsylvania-based business is the world's largest producer of tantalum powder, wire, and foil. It serves the computer, automotive, medical, defense, and consumer markets.

* * *

Because of our large holdings in natural gas, we had always been interested in liquefied natural gas. It all started in 1920, when shortly after leaving Harvard I wrote a paper on liquefying natural gas during slow periods in the summer in order to provide for the heavy winter loads. The paper was badly flawed not only because of my inexperience in cryogenics but also because I failed to foresee two very important factors: First, ordinary steel becomes very brittle at −265°F; and second, the insulation must be hermetically sealed or it will collect ice from the moisture vapor in the atmosphere and lose its insulating quality. The paper was of no interest to my father, so I gave a copy to Howell Cooper, chief engineer for Hope Natural Gas Company and other gas subsidiaries of Standard Oil. It evidently interested the Standard Oil engineers because a few years later they built, in the heart of Cleveland, a tank made of an alloy that had been tested for brittleness at very low temperatures and that had its insulation sealed from contamination from the damp air. They were confident that the tank would be able to handle the thermoshocks produced when fresh liquid methane was poured into it, but as a precautionary measure they surrounded the tank with an earthen barricade designed to stop the flow of liquid methane into the city should a flaw in the tank be exposed. Unfortunately, the contractor who built the tank decided a drain was needed to take care of rain water that collected inside the dike. The drain was connected to the city's sewers. About a year after the tank went into operation it cracked, and the liquid gas flowed into the sewers where it was ignited from an unknown source. The resulting explosion took many lives and cost many millions of dollars. It was a costly accident, and any further attempt to liquefy gas for peak loads or for transportation was set back about twenty years.

A wealthy Bostonian, Frederick Prince, hired an engineer named Morrison to advise him on his investments in the Chicago area, which gave him control of the stock yards and the meat packing firm of Armour. Prince owned some gas wells in Louisiana and was able to sell his gas for only a very small fraction of the cost of energy in Chicago. Morrison recommended that he liquefy the gas in Louisiana and ship it by barge up the Mississippi to the Chicago area, where he could use the latent cold of the gas and the heat of combustion. Morrison built a small tank and lined it with balsa wood. He was able to show that although rifle bullets could penetrate the steel tank and the wood inside, the liquid gas coming back through the path of the bullet would not cool the tank enough to make it brittle. He received permission from the Coast Guard to build a larger tank on a barge. Unfortunately, when the larger tank was built, the liquid gas shrank the balsa wood so much that there were large gaps between the blocks of wood, and the tank collapsed. Prince sold this idea to several oil companies, including Shell, and they built a small ocean-going tanker, carefully designed to cope with the problems that had caused the previous disasters. This small tanker was intended to carry a shipment of liquid natural gas from Algeria to England (this was before natural gas had been discovered in the North Sea). None of the costly liquefaction plants in Algeria had yet been completed, nor had the gas liquefaction facility in the lower Thames River. Cabot chartered the ship to move butane from our natural gasoline plants in Texas to the butadiene plant that we had built in southern France. The ship proved to be a reliable way of transporting the gas. A French company, in which we had a small interest, built much larger tankers, and there are now more than one hundred large tankers bringing gas from North Africa into European ports, from the East Indies into Japanese ports, from Alaska to Japan, and from the Persian Gulf to various Asian destinations.

By the mid-1960s we felt that we should get into this business and so we did a very thorough economic study, which indicated that we could make a reasonable profit by establishing tanks in the Boston area. In the late 1960s, we bought an option on land on the Mystic River in the city of Everett and applied for the necessary permits. Interstate transportation of gas within the United States was then under the jurisdiction of the Federal Power Commission, and we thought that we would encounter some difficulties in dealing with the commission. But the permit came through after the commission voted, in a three to two vote, that they had no jurisdiction over imports. In 1970, with all the many permits in hand, we went ahead with the project and invested about $28 million. We constructed two large tanks, the gas liquefaction facilities, and the dock, dredged and widened the river, and purchased all the necessary safety equipment.

As we became more involved in the operation, there were some scare stories in the media about what could happen if there was ever a collision in Boston Harbor and some other terrifying possibilities. We deflected these stories with expert testimony, and in 1971 we began bringing in huge tankers from Algeria every few weeks.

Before we had even finished construction in Everett, the two large gas distributors in the New York/New Jersey area begged us to construct a similar but much larger gas terminal on Staten Island. When one of the gas distributors offered $60 million in nonrecourse loans, we decided to go ahead and build the plant. We obtained the necessary permits, but before the construction really got under way, the gas distributors decided that they wanted to double the size of the enterprise. The estimate of profits from the second half looked so promising that we decided to risk our own money and we invested $35 million in the project. We also spent an additional $5 million on a large barge designed to transport the liquid between Everett and Staten Island.

When the facility in Everett had been operating for over a year and the Staten Island facility was about 80 percent complete, we received a notice from the Federal Power Commission stating that we needed a permit in order to continue our operations at Everett and Staten Island. To discontinue the construction at Staten Island was possible, but very costly. To discontinue the operation at Everett was impossible because we had a tanker at sea en route from Algiers. It could not be returned to Algiers because there was no unloading facility there, and it could not be unloaded at sea without endangering the ship and its crew. With no other place it could be sent, we were forced to go to the courts for a stay, which we did. Our lawyers thought it might take several years to get the permit to continue construction on Staten Island, so we decided to write off our investment of $35 million, sell the barge we had built, and let the creditors take over the future of the facility. The facility on Staten Island has lain idle now for twenty years, and estimates of the costs to the creditors who took over that property are about half a billion dollars—all of which should be collectable from the gas consumers of the area if the regulated gas companies are allowed to earn the normal return on their whole rate base, which would include the creditors' lost investments.

It is easy to see what had happened in the Federal Power Commission: two of the three members on the commission who had voted that the commission would have no jurisdiction over imports had retired and had not been replaced when the remaining two members reversed the decision of the commission and called for a hearing. Before the court's temporary stay expired, we were able to keep the Everett facility opened. But then a much worse problem confronted us.

The Natural Gas Act, passed by Congress in 1984, transferred the control of interstate and imported natural gas from the Federal Power Commission to the new Department of Energy. This act gave the Department of Energy the right to void all interstate contracts, although they had no jurisdiction over the contract under which we were buying gas in Algeria. By then we had negotiated contracts with fourteen utility companies to sell the gas in the New England area, all with a clause that required us to take, or pay for, a minimum fixed amount of gas, with rights to exceed this minimum when we saw the need. The contract with the Algerians was a similar, or back-to-back, contract, which required a minimum of fourteen cargoes. The Energy Department canceled our fourteen domestic contracts, leaving us dependent on current market prices in New England, which had dropped by about 50 percent. With the fixed price from Algiers at almost twice the price in New England, we were losing $5 million per cargo and we were still committed to fourteen cargoes per year. These commitments were made by a subsidiary of the parent company. The only way to get relief was to put that company through bankruptcy, but the Algerian contract contained a clause, which we had demanded, that called for arbitration and jurisdiction under English law. Escape by bankruptcy is much tougher in England, and before long we found ourselves in an attack against the parent company. Our lawyers thought that, although this might result in as much as ten years of litigation, the result would be in our favor. However, it was obvious that we could not operate with a multi-million dollar contingent liability, which might bankrupt the parent company; such a threat would make us powerless to borrow or to make any substantial commitment to engage in new enterprises. We had to find some way to deal with the Algerians. In 1987, our new chairman, Bodman, went to Algiers to see what could be done and finally settled the matter by signing a new contract that gave us freedom to not take cargoes on which we could not make a profit and gave the Algerians freedom to not ship if they found the price unacceptable. With this contract, we could go back into operation profitably. We are now back in the LNG business in Everett and are studying other LNG shipment enterprises, including ventures in Puerto Rico and Trinidad and Tobago.

* * *

Much about the businesses of today's Cabot Corporation would be recognizable to founder Godfrey Cabot. Carbon black continues to be the largest of its five specialty chemicals and materials businesses, with twenty-eight plants in operation worldwide and one billion pounds of capacity in each of the world's four regions: North America, Latin America, Pacific Asia, and Europe. We manufacture fumed silica in three countries, placing us number one in North America and number two worldwide. The plastics business, which continues to emphasize the manufacture of thermoplastic concentrates (black and white masterbatches), has plants in Belgium, Italy, the United Kingdom, and Hong Kong. A fourth specialty chemicals business, Cabot Performance Materials, is the world's largest producer of capacitor tantalum products and a major producer of niobium metal and alloys, linking it back to the Company's entry into the metals business in 1970.

Cabot Safety Corporation, which emerged from the union of our E-A-R Division and American Optical's AO Safety Division, continues to supply the world's industries with eye and hearing protection products. Our ear plugs are used by most of the defense and industrial markets, and we sell a large percentage of the safety glasses currently used, plus hard hats, goggles, face shields, respirators, damping materials, and molded foams. We also create applications for our energy-absorbing and shatter-resistant polymers.

The energy businesses, which connect the company back to its early years in oil and gas exploration and supply, include Cabot LNG Corporation and TUCO, a Texas-based coal fuel services operation that the company's Energy Group acquired in 1979. The purchase added two gas processing plants, five hundred miles of pipeline, and a coal handling business to the Company's southwest operations. Our business of drilling for oil and gas has been successful on the whole, but some enormous reserves that we discovered failed to make a profit because we were deprived of certain markets by federal regulation. Today, this part of our business has been spun off into a separate corporation, Cabot Oil & Gas Company, which was passed on to our shareholders in 1990.

Cabot has also maintained its early reputation for innovation. Today, in the specialty chemicals and materials businesses, Cabot employees are applying modern science to old technologies, creating new lives for its products by exploring the production, behavior, and handling of carbon black, fumed silica, and tantalum microaggregates and microparticulates. The results of these explorations include the discovery of specialty applications of carbon black, fumed silica, and tantalum, which add significant value to customers' end-use products.

Godfrey Cabot would also recognize the values and character of the company and its 5,400 employees. Quality, customer service, safety and environmental responsibility, cost management, teamwork, and "smart"—not just "hard"—work are practiced daily in the twenty-five countries in which the company now conducts business. The joy of discovery and the willingness to take risks are displayed in the company's research and development laboratories and in the start-up of new plants in such places as Hong Kong; Shanghai, China; Kashima, Japan; Cilegon, Indonesia; and Barry, Wales. From its international beginning in Stanlow, Cabot has extended its global presence and solidified its competitive position. Cabot Corporation looks toward the next century with excitement and the confidence that Godfrey Cabot would be proud of what has been done with his inheritance.

Robert Galvin

Communication: The Lever of Effectiveness and Productivity

O NE'S EFFECTIVENESS IN EMPLOYING the attributes of leadership and the strengths of a company is enhanced by one's aptitude for communication. When leaders and managers are better understood and appreciated, they are followed and supported. But how much better can leaders and managers be, and in what *special* ways?

While still in my twenties, I was invited to speak on a panel of three people. The topic was, how does business communicate? When it was my turn to speak, I stepped to the microphone and said the following: "Business communicates by the spoken word, the written word, and a special process called the rumor." I nodded appreciatively to the audience and began the return to my seat. The chairman was surprised and unsettled, as was the audience. Before anyone could make too much of this discomforting statement, I returned to the dais and added, "And by impression. I do not know what impression I have made, although I assume that it is certainly unfavorable to some degree. But I will suffer that if it means that you will carry away indelibly the importance of an impression as a major communications phenomenon. An unintended bad impression can be devastating to a key initiative. A good one, naturally conveyed, can carry the day."

In June 1940, I began my first paying job at Motorola. My father had founded the company twelve years earlier, and it had become a leading car radio manufacturer in addition to offering home

---

Robert Galvin is Chairman of the Executive Committee of Motorola, Inc.

radios and phonographs. As the boss's son, it was a foregone conclusion that I would assume a defined entry level position right after high school graduation.

To insure that I was on time for the introductory process, I arrived at Personnel at about 6:55 A.M. for a 7:30 A.M. start time. I took the second position on the bench directly outside of the two-man personnel department. Eddie had arrived earlier. He was a stranger then, but in the years to come he would become my friend.

The bench was just that: a painted board with wooden legs that hugged the wall, which acted as its backrest. It could not take up too much room, as the aisle where it was located was the only regular entryway into the building for all employees.

Every so often, among the arriving employees, one would acknowledge me, wave, or say hello. This must have puzzled Eddie. How could he have known that I had been in and around the place for years? Just then, Carl Holys passed by, did a double take, and turned to greet me directly. He was in charge of Inspection and was a key factory manager. He asked what I was doing there, as if he did not know. "I am here to begin my first job," I replied. "Of course, I'd heard," he said. "Come on. I will take you in to George [Lambert] to get you started." "No, thank you, Mr. Holys," I said. "This man was here before me."

Holys was used to doing whatever it took to get things done and he pressed me harder. "No, thank you," I repeated. "I will wait my turn."

I thought no more of it. The next summer, when I was well into my second summertime factory job, some of the fellows on our department softball team let it slip that by lunchtime that first day back in June 1940, the word around the plant was that the old man's son had come to work, and he was willing to wait his turn.

That was the summer between my freshman and sophomore years at Notre Dame. I had made the debate team and was teamed with another freshman, Bill Lawless, who eventually became Chief Justice of the New York Supreme Court. Bill and I had quite a run of victories, using the basics as well as every rhetorical and staging trick in the book. Our coach, himself only a graduate student, became less and less enamored with our tactics and finally sat us down for a talk. We were headed for a fall, a big one, in his evaluation, and he finally made his point in an unforgettable way: "Remember, men, the essence of a good speech is having something to say!" Our form was great; it was substance that we were lacking.

That was the single most important lesson I learned in my two successful years at Notre Dame; it shaped my speechmaking for the next fifty years. I never took a speaking assignment inside or outside the company unless I was confident that I could contribute material that was of notable value and present it in an appealing way. How often have we heard a business official elegantly recite all-too-common knowledge, concluding with a generic admonition that we had better do something about "it"? To this, knowing audiences literally shake their heads and wonder why the speaker wasted their time.

Panels, job-entry impressions, and speeches that say something new, as important as they are, are infrequent examples of communication in business. Yet, the lessons we learn from them are a prominent illustration of the effect of communication on everyday experiences. If we are late to a meeting, what message does that send? If our contribution to day-to-day deliberations is simply a rehash of unimaginative questions or ideas, what value do we add?

But I am getting ahead of myself. Communication does not start with what "I" have to say. Rather, it starts with the other person. It starts with what I hear.

My father was the boss. In fact, he was the principal owner. He could have the first and last word, but he was more interested in what others had to say. After lunch in the factory's modest cafeteria, he often would amble through the plant. He might stop and chat briefly and purposely with employees that he saw—for example, Mary Qualysa in coil winding, Rhoda Thill on the line, Katherine Wendel at the switchboard, or Joe Kubiac the carpenter. If he sensed a hint of discontent, he would follow up on this feeling. If a change was in order, even for something of his original making, he would make the change, even if it meant admitting an error.

If a picture is worth a thousand words, then the magnitude of a sensitive, prompt, objective action is even more telling. In 1947 my father and the senior officers decided that a generous deferred profit-sharing plan held more value than the 10 percent annual cash bonus that they gave. Communicating this change was going to be a challenge. A comprehensive plan, including quiet pre-announcement supervisory training leading eventually to memos, pamphlets, and posters, was prepared, with the central event to be a series of evening meetings for employees and their families. On the day of the first evening meeting, the subject was initially revealed by a simple teaser: "P. V. [my father] is going to formally announce a profit-sharing plan that will replace the bonus. The supervisors' meetings, pamphlets, etc. will follow in the next days."

My father, other seniors, and I arrived early to the evening meeting to greet the first families. They straggled in. By the appointed hour, only scores had shown rather than the multitude invited. My father was bewildered, shocked, and inwardly livid. But the show went on.

Later there was a post-meeting caucus. Was this a revolt by those who so loyally had followed him—nonunion—through the founding years? Did one of us make a mistake in issuing the invitations? In no uncertain terms, a team of us were commanded to mingle the following morning with the employees who did not attend the meeting. A full unvarnished report was to be given to my father.

The report was electrifying! From almost every person the explanation was, "I did not think I needed to come; I was already sold. If P. V. says profit sharing instead of a bonus, that is good enough for me. He has never let us down. I knew that I could learn the details later."

The report broadcasted an overriding message of credibility. Putting the new idea across took nothing but trust (nothing but trust?!). It would not have made much difference how Paul Galvin worded the justification. The year-after-year "good vibes" had set the stage for automatic agreeability. The other evening meetings were unnecessary.

The details of the plan required a clear and full explanation, but that process too went well. Because it was a contributory plan, employees had to sign up. All but one employee signed; he decided that he did not want to leave a bigger-than-expected estate to his disagreeable wife.

Customer communications must head the list of communication factors that can improve a company's effectiveness, but most leaders in most companies do poorly in this area. How can this be when they call on customers regularly?

Some years ago, an associate prompted me to visit with our customers more regularly. I decided that I would spend one day each month with a different customer. Our representatives arranged it so that I could talk in detail to the people who engineered our product into theirs and who bought, expedited, assembled, inventoried, repaired, installed, serviced, and paid our invoice. I avoided, if possible, ceremonial visits with key officials.

Our customers appreciated these visits. I followed up each visit with a detailed report, often ten to twelve pages, single-spaced, that outlined the deficiencies that I noticed. As one would assume, the improved product and service that followed sustained market share that was at risk and gained share that was pending. The extra attention, although marginal in importance, did not go unnoticed.

Two important potential consequences had yet to fully play out. First, the most evident organization/authority factor we discovered was that every deficiency I discovered and reported had been known by our sales and applications representatives for months and even years. The field people had communicated these deficiencies and cajoled their bosses, laboratories, and factories. Many of their other suggestions and complaints had been handled, but the home office and plant had priorities of their own that superseded those of the customers, causing the remaining important issues to languish.

An outstanding proposal was made that we should turn the organization chart right-side up. We should vest in each experienced, credentialed representative the authority of the Chief Executive Office to influence and effect timely responses to the customers' needs.

When we finally overcome our old bias that only supervisors can make prioritizing decisions and realize that we cannot do everything, we will come to appreciate the fact that promptly addressing each customer-service problem clears the way for more consistent customer-satisfying and market-building results. Additionally, we can cut out the redundant and irritating communication of complaints and defenses between the parties.

Second, my example was a role model to other officers, many of whom had never made a customer call. Some of them returned from their initial experience virtually born again as they integrated the customers' expectation into their departments. It is inconceivable to me that a scientist, engineer, personnel manager, senior accountant, or officer of any kind can achieve total customer satisfaction without understanding and having a feel for customers firsthand. The quality, effectiveness, and productivity of customer/supplier communication—people to people—can be doubled and even tripled.

Writing the rules is a crucial step in executing a business strategy. It calls into play critical communication credentials, expertise, and often boldness. Would-be great leaders too often fall short of positioning themselves in relation to, and thus sufficiently influencing, the more ambitious rules of the game from which to project their business offerings. Thus, a cap is placed on otherwise promising operations.

I first learned this lesson by observing the company's senior employees when I was a youth. Their travels attracted my attention. They would regularly visit the radio bureau in Washington—what is the Federal Communications Commission today—to help set standards and licensing regulations and persuade radio frequency spectrum allocations for two-way radio systems. Although all competitors could avail themselves of the same fair rules that we proposed over the years in this field and others, we tried to make sure that those provisions were the least limiting and promised the most expandability of use. To these we would then apply more exceptional applications of our products and services.

Over the years, we have applied this kind of thinking and effort to anything that could expand our opportunities. For example, we have influenced the liberalization of private enterprise in countries like Israel, France, China, and others and shaped trade policy to various nations, most notably Japan. Boldness was a frequent component of these communication campaigns.

If there ever were exceptional classes of subjects dependent on "having something to say"—substantive content—these expansion efforts more than qualified. The leaders who carry these kinds of proposals to the power source must have exceptional credentials, acquired expertise, and earned credibility. There is a strategy that can brew such qualities.

Recognizing as a youth how important rule-writing would be in the future, I began my preparations in the old neighborhood. As a young married man, I did simple chores for the property owners' association and the parish. Subchairmanships soon came my way. Then there were turns to be taken in civic fund-raising. Of course, they were always looking for the next chairman. Minor jobs for my party led to major support roles for candidates and the organization. Trade association duties led to the presidency of the association, which led to testifying on behalf of the industry before agencies and Congress. My keen interest in particular public issues like national security affairs prompted the solicitation of my opinions in ever-higher public places. All these earned Presidential Commission memberships, Advisory Committee chairmanships, and others, always in fields honorably beneficial to our business.

The table had turned. Although I was always biased to listen first, this strategy of ingratiating oneself through merit with the system's leaders figuratively "creates listeners"—influentials who are willing to be influenced. In the final analysis, if you wish to be heard, it is best to cause others to want to hear you.

The most dramatic demonstration of this is our experience resulting from being the first large company to be awarded the Malcolm Baldrige National Quality Award in 1988. We had purposefully and steadily improved our quality processes and results since commencing a recommitment to quality in 1979. When the announcement of the Award program was made in 1987, we dared to nominate ourselves, hoping that we were in step with the best or perhaps even a step ahead.

When you win, you are expected to share ideas with others. It turned out that the interest from other companies was virtually insatiable, and so we were willing to share information liberally with any responsible organization, be it customer, competitor, supplier, school, or government agency. It further transpired that what we had to offer was sufficiently useful and unique so that the interest was sustained.

The privilege that this accorded us was overwhelming. How often had we wished that a prospective customer would give us ten minutes to present ourselves? As a result of the Baldrige program, thousands of institutions invited us to demonstrate the substantive qualities of our processes, practices, policies, and values, allowing us in the process to describe our products and services.

There are two remaining communication opportunities that can increase a company's effectiveness and productivity. Does advertising produce economic growth? Yes, particularly if it simply and clearly conveys a message. Everything else is superfluous, and those "other things" often mask the real meaning intended by the advertiser. A worthy ad must mostly pronounce news that benefits you, the reader or listener. That is the test.

But the more important voice is the greater voice of our people. They see. They know. They deserve to be heard more. Most of our companies have employed hundreds of ways to encourage in-volvement, and "teams" now heads the list.

We have been pursuing some methods of improvement since the late 1940s. Participation and participation management have increasingly added value and enhanced employee and customer satisfaction over the years. But the general effectiveness has been augmented recently by the adoption of what we refer to as "team competition."

Today, over half of our employees from around the world are on problem-solving teams, comprised of a dozen or so associates, that compete annually in a ladder-type tournament. Some teams are single-function in composition; many are staffed cross-functionally, and suppliers or customers may also be members. They are judged on the criteria of demonstrable problem-solving skills and demon-strable results.

To prepare, each participant must be solidly educated in a raft of talents such as issue identity, data acquisition, statistical analysis, diagramming problems, creativity techniques, group dynamics, communications skills, and, yes, public presentation. The "com-munications through education" content for most team mem-bers is an enhanced study of many of these subjects again, at a more "graduate" level.

Each team deals with many problems but focuses on one in particular because of its practical importance. Each team presents its experience and the result of its significant work in a twelve-minute stage presentation to a large audience of peers and officials with as much fanfare as the team wishes. Winners advance to regional and international finals. The judging factors include an evaluation of the transfer of proposed solutions across the com-pany to relevant company organizations who can learn and im-prove from the other's better method or idea. Consequently, at pivotal times during phases of the competition period, the volume of effective communication exchange is literally equal to the capac-ity of a high-power computer. The motivation is boundless. Inciden-tally, some teams prefer not to compete, and that is acceptable. Their achievements often challenge in utility the competitor-types. Through it all, the company's regular work gets done.

The issues and problems addressed cover everything. Savings in time and money in regard to the chosen subject are universal, and these range from moderate to a virtual pot of gold. Above all, each generates an improvement in quality of 30 to 80 percent. Total customer satisfaction is overriding.

We are no longer unduly surprised when a team right out of the ranks comes forth with a profound resolution of inestimable value. For example, the employees who operate the cafeteria in our Penang, Malaysia plant were aware of the team activity and competition but had not formally enrolled. They had a customer satisfaction problem concerning the time it took for each diner to proceed through food service and check out so that the patron had time to enjoy the remaining lunch period.

The cafeteria personnel consisted of eight people, including a cook, food handlers, cashiers, and dishwashers, and not one had an advanced education. But they huddled together, seriously and frequently, to discuss some of the techniques they had overheard. Even more importantly, they applied common sense.

They relocated the cups to avoid congestion and simplified the dishing of food to reduce time at that station. But these timesaving improvements were marginal at best. The checkout counters were an obvious bottleneck. How about arranging for more cashiers through job reassignments? That helped. Why not have one checkout line for customers with exact change? That worked. Why not remove the cashier from the checkout line entirely and allow the diner with exact change to simply deposit it into a container? Now they were getting somewhere. Finally, one of them suggested, why don't we eliminate the cashier function entirely? Place baskets of change conveniently in the eating area. Have the diners deposit their payment at their convenience and if they do not have the exact coinage, let them reach in and make their own. "Why don't we trust our customers?" Trust! Wow! That is profound.

The time issue evaporated, of course, but that became secondary to the plant-wide elevation of self-esteem. The whole experience bespoke the good words, "we are trusted." What a message! What a productivity energizer!

Special communication opportunities abound. They act like levers, raising performance continually and substantially. The only problem is that the leader personally has to key the phenomena. The special communication opportunities mentioned here must be personally role modeled, initiated, and supported by a company's leaders. Fortunately, they are not all that hard or that many.

Stephen D. Bechtel, Jr.

Reflections on Success

INTRODUCTION

A SKED TO COMMENT ON WHAT LED to my personal success and the success of Bechtel Group, Inc., and what the future might bring for business, I respond by first noting how success is often described: Corporate success is viewed by many in terms of market domination, leadership, profitability, and growth in equity value. An individual's business success is viewed in terms of "personal power," eminence, titles, and compensation.

For me, these definitions are only part of the story and too often are incomplete or distorted criteria for such evaluations. I believe the most important measure of "business success" should be personal satisfaction from the constructive accomplishments of an individual and/or a team.

MY PERSONAL SUCCESS

As a child and young adult, the exposure I had to my parents' values and conduct and to their friends and associates was extremely valuable. They imprinted on me fundamentals that have endured throughout my life.

I was privileged, in our home and in visiting Bechtel's offices and jobs with my father, to be able to observe Bechtel's people and other leading engineers, contractors, and clients. I had the chance to sit in on many business discussions. Without question, these

Stephen D. Bechtel, Jr. is Chairman Emeritus of Bechtel Group, Inc., an international engineering and construction company, and Fremont Group, Inc., a private investment company.

169

opportunities had much to do with the shaping of my own values and with the way I handled situations and dealt with people. I learned a lot just by listening; often I was not conscious of the lessons I was learning. By osmosis, I was absorbing personal values, thinking processes, ways of talking, and ways of doing things.

I was fortunate, too, in my formal education. From junior high school and high school, I particularly remember my math and science classes, mechanical drawing, grammar, and shop. These courses impressed me positively and deeply, and they gave me the foundation I needed to benefit from more advanced education.

While not particularly good at varsity sports, I played "B" basketball and earned a varsity letter as manager of the football team, learning that you did not actually have to play the game to get a letter. Being a Boy Scout was another inspirational childhood experience. As a member of a good troop with capable leadership, I took seriously the lessons and teachings of scouting. The scout oath and laws (as well as the tests, merit badges, and outdoor camping) helped clarify and confirm my personal values and beliefs.

Working while attending high school taught me the value of a dollar and gave me a sense of responsibility at a young age. My first job, at age fifteen, was working in a machine shop as a sweeper (wages were fifty cents an hour). Eventually I was allowed to run a drill press. The following summer, I got my first construction job as a "stake-puncher" on a survey crew. During the summer and on the weekends of my senior year of high school, I worked on a survey crew that was involved in the building of a shipyard and the first ship in the yard.

I discovered sailing during this period, and for several years I was involved in some very competitive sailboat racing. Besides the actual sailing and competition, I did much of the wood and metal work, transforming the boats from bare hulls into finished vessels. I also took care of the maintenance of the boats. I raced all over California and traveled one summer to Texas, where I won the Junior National Snipe Championship and placed second in the International Snipe Championship. I attribute much of my subsequent competitive nature and ability to function well under pressure to my sailing experiences. The saying "steady at the helm" is good advice, not only in competitive sailing but in many other life situations.

US Marine Corps and College Education

In my last year of high school, during World War II, I enlisted in the Marine Corps Reserve. Called to active duty upon graduation from high school, I was sent off to college within a few days as part of the officer training program. The combination of college and Marine Corps training was a great learning and shaping experience. I had to contend with not only a high-powered academic program, but also the "spit and polish" and physical fitness of the Marine Corps. The officer in charge of our detachment was very good but very tough, as were the NCOs.

I enrolled in an accelerated program for my engineering degree, completing four academic years in two and two-thirds calendar years. An efficient use of time was one of the many things I learned from the combination of active duty in the Marine Corps Reserve and undergraduate engineering courses. We woke up early in the morning to do calisthenics and road work before class, with room inspection every morning and lights out at 10:00 P.M. We worked under a very pressing incentive arrangement: if our academic grades and general conduct were not satisfactory, we would be removed from the officer training program and sent to boot camp, where there was every possibility that we would be sent overseas to the Pacific Islands into some of the most difficult campaigns.

I learned a tremendous amount from some very inspiring teachers in my math, physics, and civil engineering courses. The most memorable was Charlie Ellis, a structures professor at Purdue. He did not let his students use textbook formulas; we had to derive all of the solutions to our structural problems from basic mechanics. This approach taught me how to think a problem through, relying only on the fundamentals.

Following World War II and my marriage, I enrolled in Stanford's Graduate School of Business. The normal two-year MBA program was compressed into one and a half calendar years. It was the first regular class after World War II, and I was in the company of some very high-powered students, many of whom were much older. The competition was fierce. From courses dealing with organization, finance, marketing, accounting, and business law, I gained a great deal of knowledge that has proven very useful to me in exercising leadership and judgment throughout my business career.

I was blessed with some outstanding and very inspirational teachers at Stanford, including Paul Holden, Dave Faville, Bud Petersen, and Knight Allen. I was also in the company of some outstanding schoolmates: Frank Cary, who became CEO of IBM; Don Peterson, who went on to become CEO of Ford Motor Company; and Chuck Robinson, who headed Marcona Mining and was Under Secretary of State at one point.

My active duty in the Marine Corps Reserve, my engineering education, and my graduate work gave me a wonderful and very valuable foundation for my later business activities and responsibilities.

Early Work Experience: Pipelining

While growing up, I was undecided about whether I would join the family business or go out on my own. I knew I wanted to go into construction work, but I recognized some disadvantages of going into the family business.

A turning point came, however, when my wife and I had the chance to make a trip around the world with my mother and father upon my completion of graduate school. We visited first the Far East, then the Middle East, where Bechtel had a lot of work going on. En route back home, I heard about a new pipeline contract just awarded to Bechtel, and it sparked my interest.

I began working under one of Bechtel's senior officers and went to Texas as a "field engineer." My assignment included doing whatever field and cost engineering was needed, plus driving a low-bed truck part-time at night. The job required us to work seven days a week, ten hours per day, but most of us put in fourteen to sixteen hours a day. It was very demanding but fascinating work.

My second assignment was on the construction of a major 34-inch pipeline in California, the first of its size ever built. I was the Job Field Engineer, and one of my responsibilities was to help the Master Mechanic develop and keep the pipeline construction equipment working. Initially it was not heavy or large enough to handle the 34-inch pipe. We had some very interesting times, to say the least.

I then moved on to become Assistant Superintendent, and my responsibilities were varied. I was in charge of a "dirt moving spread," building the roads and right-of-way while excavating the ditch across the Tehachapi Mountains for the same pipeline. It was very challenging and satisfying to be in charge of the operation, especially since my boss was several hundred miles away looking after the big pipeline spread.

Later, I was made Spread Superintendent and was put on a new pipeline contract that the company had in southern Illinois, near East St. Louis. This project involved an extremely difficult union labor situation. To complicate matters, because of a steel strike, we ended up laying the pipeline in winter under very adverse weather conditions. However, we completed the work in reasonably good order and made money on the job.

These successes gave me the confidence to know that I could handle people in difficult situations and, in the vernacular of construction, run a good, "hard-money" job. Shortly afterward, I was elected Vice President of Canadian Bechtel Limited and moved to British Columbia. I worked with the Project Manager, who represented the owner, to draw up, take bids for, and recommend the awarding of contracts for the construction of a 700-mile oil pipeline, the first across the Canadian Rockies. I later moved back to San Francisco and was appointed Division Manager for all of our pipeline work.

Pipelining is a unique type of construction, and pipeliners are a breed all their own. It is very risky work, fast-paced, with great financial penalties and rewards depending on how the job goes. I learned a tremendous amount from pipeline work and from the people associated with it. They were hard-working, fun-loving people who had a strong work ethic.

Throughout my early work experiences, I had the opportunity to work for, with, and around some truly great people. Few experiences in life are as beneficial to a young man as working for effective, diligent, top-caliber bosses.

Becoming a Licensed Engineer

About six or eight years after completion of my university education, I was asked by my boss to become licensed as an engineer in the state of New York, where our company had legal problems using the Bechtel name. My father, who was then president of the company, was not and could not be licensed there. I had to sit for a twelve-hour written examination that covered nearly all of the subjects I had taken to earn my engineering degree. Simultaneously, I was Division Manager of the Pipeline Division and had some very consuming responsibilities. However, with the help of some very good tutors (and with the shared knowledge of everyone at Bechtel), I took the examination, passed, and became licensed. I later had to be licensed in several states, three of which did not accept by reciprocity the New York license. Michigan, California, and Alaska required me to take additional exams. While it was a great distraction at the time, brushing up on all of the subjects and the testing were a healthy challenge for me.

Transition to Senior Management

Shortly after I became Pipeline Division Manager, I was given the additional responsibility of being Treasurer of the Corporation. This assignment was interesting and very challenging. In the course of fulfilling these responsibilities, I developed and set up a system of projecting and forecasting revenues, costs, overhead, and profits. Until then, our company had never really forecasted new work to be booked or the financial outcome of calendar years. The fundamentals of this system continue today as a primary tool used for controlling our cash, our costs, and the planning of staff levels.

For a few years following the completion of my pipeline work in the field, there were many changes in our business. Besides Pipeline and Treasury responsibilities, I was responsible for the International Division and the Equipment Department and was elected Director and then Chairman of the newly established Executive Committee.

In 1960, at the age of thirty-five, I became President of the company. My father, who was then sixty years old, shifted his title to Chairman. By agreement, there was no CEO; I was to run the business with his support. I enjoyed working for and with my father in this arrangement for five great years.

In taking on the leadership of our business, I gave serious thought to the priorities of the business at the time, and I tried to anticipate what its future needs would be. The business was doing extremely well and the outlook was very good. However, I have always had a personal philosophy of trying to improve both my own performance and the performance of the operation in which I was involved. I recognized a natural tendency for a "new broom" to want to sweep things clean and make its own mark. However, because things were going so well and the outlook was so good, I decided that my approach to change was going to be by evolution, not revolution. This approach served me well over the next decade as I earned the support of the organization and developed more confidence in my own assessment of new initiatives and the best way to accomplish them.

I owe most of my success during the thirty years that I served as President and Chairman to the outstanding individuals who made up the great team working with and for me. The business grew from a non-manual staff of 3,900 in 1960 to 44,500 at the peak in 1982. Revenue increased from $460 million in 1960 to $14.3 billion in 1983. Then, the available contracts dropped off drastically, and we downsized the organization to a staff level of 17,400 in 1987, with revenue of $4.5 billion. The business has subsequently been built back up; at the time of my retirement in 1990, we had a staff of 21,400. At year-end 1994 we had a staff of 17,000 employees and annual revenue of $7.8 billion.

Successfully managing alternating periods of growth and downsizing is a challenge to which few are able to rise. It is critical to recognize realities quickly, and it is extremely important for not only top management but for the entire organization to respond flexibly and quickly to conditions as they develop. Future market conditions are never certain. Quite often in turbulent times, they are even less clear, with contradicting indices. The answer is simple: you must be on top of changing conditions and be able to act appropriately.

I have a strong belief that there are more ways of doing business than just across a desk. In our industry, we need to understand our client's business, his problems, and the personalities involved. These understandings, often developed through regular "in-office" business dealings, can be greatly enhanced with social and recreational activities that help to build understanding and trust, enabling you to better anticipate your client's needs.

I also gained a wealth of knowledge and judgment through service on some outside corporate boards, such as the Tennessee Gas Transmission Company, General Motors, and International Business Machines Corporation, and from being involved with other industry activities including the Conference Board, the Policy Committee of The Business Roundtable, and the Business Council. In addition, I was on the Board of Trustees of the California Institute of Technology and have chaired the National Academy of Engineering. All of these gave me valuable insight and new perspectives about general business problems, approaches to running a business, and new and emerging technologies.

Family Support

My wife, Betty, and I were married shortly after I completed my time in the Marine Corps and had finished my undergraduate engineering studies. She helped me get through graduate school and has been a source of strength and support throughout my business career. I consider her a trusted friend and advisor; she is my confidante and sometimes my critic. Betty's help has been particularly important in my relationships with key Bechtel people and with our clients. Like my mother, she became a role model for other Bechtel wives. I could not have accomplished all that I have in my business career without her involvement and support.

Our five children, now grown with families of their own, have also been very understanding and supportive. In fact, one son and two sons-in-law are currently part of the top management team of our business.

Learning from Mistakes

While everyone makes mistakes, I have been very lucky. Although the company and I have made very few major mistakes, the ones that we have made have allowed us to learn some very important lessons. For instance, from our bad investments in non-engineering and construction ventures, namely our acquisition of approximately a 50 percent interest in WellTech, an oil well workover and completion company, we learned lessons about timing and over-leveraging. Our timing on WellTech was very bad. We bought this company at a normal acquisition premium over current earnings and market value when the future seemed promising. The price of oil was predicted to increase between $50 and $100 per barrel, and we anticipated a lot of oil field activity in the United States. Our forecast of the future was wrong. Further, based on our optimism, we over-leveraged the business and it assumed much more debt than it should have. Overall, the results of this investment were disastrous.

I realize now that when our engineering-construction markets declined in 1982, I should have been more forceful in directing members of the top management team to take positive action to reduce staff commensurate with the decrease in our work load.

While Bechtel "kept its head above water," we should have anticipated sooner and more accurately the extent of the downturn and responded more quickly.

Some of the valuable lessons we learned that might benefit others who are facing a weak market outlook and a potential need to downsize include: staying tuned to market changes and anticipating trends; staying tuned to competitors' activities; determining special market strategies and tactics to use if a weak market continues; alerting employees to market conditions; preparing management for necessary actions, i.e., forecasting probable staffing requirements and possible excesses; recognizing the importance of staff reduction at all levels, not just at the bottom; determining layoff policies and practices and special arrangements (namely, lead time on preliminary notices and final termination notices, selection of likely individuals for layoff, early retirement benefits, consulting and outplacement arrangements, elimination or reduction of some benefits, and granting of some merit and promotion increases); letting the "A" players know that they are the most secure; restructuring the organization as necessary; and refraining from making statements to the organization that may not be true (e.g., "We are now at the bottom," "The outlook is improving," and "We anticipate no more layoffs.").

Bechtel saw the market start to shrink in 1981–1982. It hit many of our competitors sooner than this. At first we thought that things would not prove to be that bad for us because of our diversity and the expectation that our nuclear power work could carry us reasonably well. We had gone through several down markets in one or more industry segments, but never as severely or as broadly as what we experienced in the period from 1982 to 1987. While we were a little "behind the curve" in responding to the reduced work load, we eventually downsized the staff and reduced other costs. As a result, we kept our business "in the black" cash-wise and at the bottom line all through those tough years.

Another lesson was that we should have paid more attention much sooner to better articulating our corporate culture to our people. In 1966, we formulated a statement of "Bechtel's Purposes and Objectives," which was disseminated throughout the company. Our mission, vision, and culture were then well understood by all of our people. Our employees seemed to perceive what we were all about while the business was still growing.

However, starting in 1982, when our market began a major downturn and we had to downsize with extensive layoffs, many of our people became greatly concerned about what was happening and where we were going. As a result, morale deteriorated substantially, although perhaps not as much as that of others in our industry. In 1987, with the help of our younger senior management, we began a major effort to better communicate our corporate culture to our people. With input from all levels of the organization, we developed and issued a more in-depth written statement of Bechtel's Mission and Vision. This statement, which described how Bechtel would operate as a market-driven company, expressed our philosophy of management and our goals and values.

In 1992, under the leadership of one of my sons, Riley Bechtel, a new communication effort commenced with his "Toward 2001 Strategy," which explained Bechtel's vision for where we wanted to be in the year 2001 and sketched out our plans for getting there. In addition to focusing again on our purpose, mission, and core principles, it articulated management's thinking about how we would achieve excellence in each of the following areas: maintaining our high standard of performance; marketing and selling our special strengths; innovating to add new value; being the preferred engineering and construction employer; being global; improving management and leadership; and understanding and applying technology.

"Continuous Improvement" (CI) is the methodology underlying these strategies. Although not an entirely new concept, Bechtel's "CI" process is now a concentrated effort being integrated into all line functions. It is being led by members of our senior management and involves all levels of our organization.

Leadership is recognized as being key to Bechtel's future. As part of our continuous improvement effort, Bechtel's Leadership Model—a vision of what we want our leaders to be, what we want them to know, and what we want them to do—was developed and is currently being taught to our people in seminars throughout the company.

OUR COMPANY'S SUCCESS

To be successful in the construction business is very difficult, largely because it is an extremely competitive industry. The casualty rate over the years has been high in comparison with other industries, and continued success beyond one generation of leadership is rare. A clear distinction needs to be made between short-term or one-time success and long-term, continuous success. Company success should be viewed over a long period of time, measured not only in terms of financial success (i.e., profitability and growth in value of equity) but also in terms of industry leadership, size and continued growth, reputation for quality performance, quality and cost of facilities, ethical conduct, customer relationships, and being the "employer of choice" from the viewpoint of people in the industry.

There are a variety of factors that can lead to continued success in the construction industry; what follows are the more important ones.

Leadership

Bechtel's Leadership Model states, "Today's effective leaders influence others to transcend their own self-interest for the good of the organization. They focus their team's full attention on adding value to their customers; catalyze the energies and abilities of others by creating and communicating a compelling vision of the future; inspire a sense of shared ownership among team members; develop trust through integrity and personal example; and show high standards of performance and accomplishment and motivate others to adopt such standards.

"By empowering team members, these leaders develop leadership ability in others. Thus, they ensure that their team's achievements become self-perpetuating. Rather than seeing leadership as rank and privilege, they see it as responsibility—both for the development of colleagues and the successful performance of the team's mission."

The individual attributes and judgment qualities that Bechtel considers important in a leader are integrity, technical and management competence, self-esteem, personal confidence and discipline, tolerance of ambiguity and uncertainty, resilience, a broad perspective, commitment, a willingness to take risks, persistence, self-motivation, and respect for others.

High performance teams run on trust. They achieve synergy, that is, team members work together to accomplish things they could not achieve working individually. To develop such teams, leaders must apply specific skills and competencies; these are also used in evaluating employee development, succession planning, promotions, and layoffs. These skills and competencies include coaching, interpersonal and organizational communication, empowering and motivating others, problem-solving, decision-making, teamwork, planning and organizing, a performance orientation, and mentoring.

The Caliber of our Employees

The engineering-construction business is truly a "people" business; it is a service business. While fixed assets (construction equipment and office facilities) have the potential to be an important factor, they need not be. Management of risks and expenditures and performance of work are the keys to the business. These elements are managed and performed by people. Thus, the fundamentals of a business revolve around the people in the organization and their performance. The caliber of these people, their knowledge, and their working relationships are the biggest differentiators among competitors in our industry.

I believe Bechtel employs competent and committed people. Bechtel's policy is to treat its employees fairly and to compensate them well. Bechtel shares the financial rewards of its success by providing differentiated bonuses to the top 40 percent of its non-manual staff. Further, the senior officers participate in the ownership of the business.

Values and Culture in our Business

Values, culture, beliefs, philosophy, and principles should set the direction, the conduct, and the conscience of a business organization. At Bechtel, we have always articulated our beliefs and principles to our employees. Although the wording of these beliefs and principles has been improved over the years, the basic content has remained the same.

These core beliefs and principles for Bechtel are as follows: We bring to our work a proud heritage of accomplishment, integrity, excellence, and commitment to our customers' interests, along with a willingness to appropriately adapt ourselves to change while maintaining our fundamental values and constancy of purpose. We will continue to adhere to the highest standards of ethics and integrity, understand and strive to exceed our customers' expectations, and deliver exceptional value to our customers, helping them to maximize their success. We will do our work safely, consistent with responsible environmental principles; make continuous improvement an integral part of the way we operate; attract, develop, motivate, and retain highly competent, committed, and creative colleagues of diverse origins who represent the best in their fields; promote from within as much and as early as possible; create a work environment supported by leadership that fosters openness, trust, communication, teamwork, empowerment, innovation, and satisfaction; respond to our rapidly changing world with entrepreneurial approaches, innovative solutions, advanced technology, and high-quality, timely decision-making; remain privately owned, financially prudent, and global, with ownership held by active senior management; and reward those contributing to our success.

The overarching link of these principles is a commitment to "excellence in our performance," with continuous improvement as the focus and constant goal.

Other Factors

There are other factors that are important to our company's success. They include: *1)* A knowledge of the engineering-construction industry; *2)* Good business practices, such as selecting and handling people; organizing, monitoring, and controlling; marketing and business development; finance, cost, and pricing; planning and scheduling; analyzing and managing risk; and applying technology; *3)* Focusing on the market and the customer, since the construction market is continually changing. Segments and individual customers grow and decline, sometimes cyclically, sometimes permanently. Some geographic regions grow, some stagnate, and some collapse. Competition, in the main, is always tough, but individual organizations are not always consistent in their approach to the market. Thus, for continued success, it is imperative to focus clearly on the market and customer wants and needs. *4)* Geographic and industry diversification. One of Bechtel's great strengths has been its geographic and industry diversification. The fact that we operate not only in the United States but also in many other countries and that we serve many different industries and governmental agencies helps us to weather turbulent economic swings that have seriously hurt many of our competitors. *5)* Balanced decentralization of authority, recognizing that many of our businesses differ from others. We apply only that level of standardized processes and procedure that is effective and needed. *6)* Private active ownership. Bechtel does not have "absentee" owners; it has active family and senior management ownership. This concept has been a very important factor in our continued success. Besides providing a major financial incentive to senior management, private active ownership eliminates the substantial distraction and costs of dealing with the Securities and Exchange Commission, the stock exchanges, security analysts, and dissident, uninformed shareholders. This arrangement also provides for much quicker and more thoughtful shareholder actions as they are needed.

New Considerations

The accelerating rate of change and major new technological developments during the last three decades have had a great impact on the engineering-construction business. Some business lines expanded and some new ones emerged, e.g., nuclear power, liquefied natural gas, transportation, space and defense, environmental protection and cleanup. We were fortunate that in dealing with the needs of our civil, process, and nuclear work over the years, we developed a great capacity to handle both the technical and practical aspects of air, ground, water, and sound pollution. As environmental protection and cleanup became more important, we were in a strong position to assist our customers with environmental concerns about their facilities. Our operations were greatly impacted by new developments in computers and information technologies, communications, and new materials. We have responded by being extremely alert to changing markets and new technologies as they evolve and by educating and positioning ourselves to take advantage of them.

Increased environmental concerns voiced by the public and the government caused a substantial increase in the cost and the amount of time involved in designing and building most new facilities. As a business, our approach has been three-pronged: We have participated in educating the public and appropriate government officials about the cost/benefit effects of protection as well as available remedial options, worked in the appropriate arenas to try to get reasonable standards and procedures adopted and administered, and professionally complied on our customers' behalf with the established laws and regulations. Government regulations concerning safety on work sites and licensing of nuclear power plants have also increased; we have recognized these changing requirements realistically and have accordingly worked to meet them.

In response to this change in attitude from "let the buyer beware" to "consumer protectionism," with increased liabilities in both scope and size for personal and corporate actions, we have improved the clarity of the appropriate terms in our contracts. Refusing to accept certain risks or exposures and even sometimes some assignments, we have trained our people to better manage our risks and exposure and have increased both our in-house and outside legal staff.

The Role of the Government

With the benefit of hindsight, I recognize that my "penchant for privacy" as an individual, along with the privately owned status of our company, made us targets for speculation—we were accused of being secretive. Being more open and accessible earlier in our history might have alleviated some adverse publicity that we experienced.

Employing former government officials added an outside perspective to the leadership and management expertise of our senior management team. They were not hired to represent Bechtel with the US government; in fact, by virtue of their previous work for the government, they brought added attention and scrutiny. However, on the whole, the positive benefits outweighed the negatives. Their judgments and capabilities were valuable additions to our business.

In the last five years (1990–1994), less than 20 percent of our revenue was from the US government and its agencies; the US government has not had a major role in the success of our business. However, the fact that the government's agencies (federal, state, and local) have increased their involvement in permitting and overseeing the construction of most new facilities over the last thirty or so years has probably helped us competitively in a significant way. We were able to help our customers deal with the agencies that claimed jurisdiction, whereas some of our competitors did not have the same capability. Moreover, we "contained," or limited, within our organization a lot of the bureaucratic "virus" inherent in government procurement and contracting regulations by establishing a new, separately managed, autonomous corporation that handles nearly all of our US government work.

Evolution of Management Philosophy

An organization of able, active human beings will inevitably have conflicts of style and viewpoints. Managing such a group requires good judgment, intelligent decisions, and compromises. If managed properly, the output of the organization will be significantly greater than the sum of the output of the individuals: one plus one can equal more than two. A group needs to be organized so that the individuals complement one another.

Our organizational concept has moved from an era where individuals were held responsible and accountable for the profitability of their own business lines, which created internal competition, to a team approach to doing business and delegating responsibility, which resulted in more cooperation, cohesiveness, and focus on what is good for the whole company. Although this team approach lessens the clarity of responsibility and authority, senior management is convinced that the results are better overall. The best answer is a proper balance of individual responsibility and authority with a strong emphasis on cooperation and teamwork.

Top Management Transitions

Upon the death in June 1989 of Alden Yates, then President of Bechtel Group, Inc., Riley Bechtel was elected President and Chief Operating Officer. The following year—when I reached our normal retirement age of sixty-five—I stepped down as Chairman and CEO, and Riley was elected Chief Executive Officer. The position of Chairman was left vacant. I was given the honorary title of Chairman Emeritus and continued as a member of the Board of Directors. In 1995 Riley was elected Chairman, while continuing to serve as CEO.

Prior to that time, Riley had come up through the ranks, with several field assignments in construction and business development in power, petroleum, and chemical work in the United States and overseas. After serving as Managing Director of our company in the United Kingdom, he was moved back to San Francisco as Executive Vice President in charge of all of the company's central support services. Riley assumed the presidency with great support from our top management and, to the best of my knowledge, from the total organization. Since then, he has moved very thoughtfully and positively to provide strong leadership to the company. Together with five Executive Vice Presidents, who had been with us for many years, they make an outstanding top management team.

THE BUSINESS WORLD OF THE FUTURE

Despite many uncertainties, some trends appear evident to me, in particular, the globalization of industry. Technology will advance rapidly, especially with new end products, manufacturing and processing technologies, information and communications technologies, and new materials. Local political instabilities will continue in many countries. In the United States and most, if not all, of the world, political systems will continue to be generally ineffective in providing a climate for optimum public welfare and economic development; however, as in the past, I believe that enterprising people and effective business organizations will prosper and do important, constructive work.

If asked by the leadership of our country or any other country to respond to the question of what role government should play in business and the economy, I would simply state the following: The "private sector," with free and open markets, creates the wealth of a country and provides the wherewithal for people to improve their lot and enjoy a good standard of living. There are many examples of failures of other economic systems, including socialism, communism, and other centrally planned and controlled economies.

The government's proper role is to: *1)* provide law, order, public safety, national security, foreign economic and social relations, and minimum care and support for those citizens who are unable to care for themselves; *2)* maintain conditions for free and open markets, e.g., antimonopoly legislation; and *3)* prevent "market failure," where private markets do not consider social benefits and costs, namely environmental protection, product safety and information, protection of "heritage and culture" (national parks, museums, etc.), provision of selected public infrastructure (where private incentives are inadequate to meet social needs), public education, and essential worker protection such as child labor laws.

As has happened in the United States and many other democracies, governments that go much beyond this can easily lead too many people to believe that "the government owes them a living." Such a philosophy undercuts the incentive for the individual to take responsibility for his own welfare. Eventually it will lead to the deterioration of the moral character of the populace.

The problem with many government functions is not *whether* they should be done but *how* they should be done. One principle is to create market-like incentives instead of ones that command and control, for example, providing "tradeable" pollution permits instead of arbitrarily applying the same rules for everyone to meet environmental goals; installing "user fees" to meet a significant proportion of the costs for parks, infrastructure, etc.; meeting public goals using methods of private-sector efficiency, which minimize bureaucracy; and weighing social benefits and social costs carefully. In raising finances to pay for government activity, both efficiency and equity must be taken into account, that is, incentive and disincentive effects of tax policy and "fairness" (who bears the burden of taxes).

Everything considered, I believe there will be great opportunities for business in the future, in the United States and in many places around the world. There will be a high rate of change in the globalization of information and the education and enlightenment of people around the world. Travel and communications will be easier and faster. Free markets will be expanding. Money will be more fungible, and technology will advance more rapidly.

I believe the United States, despite its many governmental bureaucracies and its short-sighted, politically motivated decision-making, will continue to be the most attractive country in the world in which to live, work, and raise families. It will be the most attractive country in the world for most types of business investment, and for many years it will continue to be the most attractive market for most products.

TO BE SUCCESSFUL IN THE FUTURE

For a business to be successful in the future, in general, a focus on fundamentals is of prime importance. A business should prepare for the future by providing a strong market and customer focus (i.e., serving the market well today and anticipating future market conditions and opportunities), strong leadership, top quality people and facilities, sound culture and values, and by being flexible and adaptable.

To be personally successful, as I define "success," I believe one should: *1)* have an outstanding character; *2)* continuously strive to improve your personal performance; *3)* be a team player; *4)* be a positive, constructive influence, and be involved in activities around you, both inside your company and in your community; 5) be open-minded, objective, and realistic—accept change as a reality, recognizing that it offers opportunities; 6) be a visionary—focus on areas where experience and abilities can be matched by few others, strive to foresee the industries and geographic areas that will offer the greatest opportunities for long-term profit, develop a competitive, innovative mentality, and create something new, uniquely suited to your company's strengths; 7) be a hard-working participant; and 8) enjoy your work and show your enthusiasm for it. It will be infectious to those around you.

I would advise someone starting up a new business to learn the fundamentals of the particular business before striking out on your own. Be prudent and sensible, sticking to the basics of good business management—develop a sound organizational structure, delegate authority along with responsibility, and set up good communications. Never be satisfied with the status quo, for you cannot run on momentum; if you try to base your future on your past, you will fall behind. Determine your company's purpose and objectives, its mission and vision, both short-term and long-term. Recognize these objectives but also the fact that your vision may change as your company evolves. Strive to be the best, not necessarily the biggest; the growth of an organization should be viewed as a growth in intellect, judgment, and knowledge, in the capabilities of individuals and the whole team. Provide your people with growth opportunities. Hire outstanding individuals and put them together into a strong, cohesive team. Recognize that teamwork does not divide responsibility but shares it and in the end allows common people to do uncommon things.

CLOSING

I have outlined some of the most important elements in my own past success and that of Bechtel Group, Inc. I believe the majority, if not all, of these factors can apply to most businesses and individuals. In the future, I believe there will be opportunities for great successes for both individuals and companies, in the United States and elsewhere, as most of the nations of the world become more interdependent, as people become more enlightened, and as technology advances.

Edward C. Johnson 3d

Adventures of a Contrarian

I F MY FATHER, who founded Fidelity Investments in 1946, could see the company today, he might say that we are too big and try to do too many things. But I am sure he would marvel at what we have become.

When he retired in 1974, the company was managing about $2 billion in fourteen funds; twenty years later, we were managing $262 billion in 209 funds and had total customer assets of $380 billion. During that time, we also added many appendages to our basic investment business, ranging from a car service to a chain of art galleries to a telecommunications company in London.

All of this growth certainly has continued to challenge us, which would have delighted my father. He believed that being too secure led to trouble. From the firm's earliest years, he encouraged us to oppose orthodox thinking. This philosophy has given us the freedom to try out new ideas, learn from our mistakes, and build on our successes; it is deeply ingrained in me and in Fidelity today.

STARTING OUT

My father was fascinated by the stock market and had a natural talent for investing. He believed that personal success came from doing what you enjoy and what you are good at, and he counseled me to pursue a career in whatever interested me most, just as he had done.

I was not sure if I would find the investment business interesting. My father had given me a healthy respect for the market—

189

a respect that came from his own experience watching a whole generation lose money in the late 1920s and 1930s. As a child, I knew you were not to play with the market, in the same way I knew not to play with matches, unless you knew what you were doing.

This is a business in which you come to know what you are doing either by buying and selling stocks on your own or buying and selling them professionally; there is no substitute for experience. I had no firsthand experience until I came to work for Fidelity in 1957, first as an analyst and later as a portfolio manager. My father believed that managers should have the freedom to take a certain amount of risk, so we learned mostly through trial and error. Some of the best money managers I have known— my father, Gerry Tsai, and Peter Lynch—developed a unique investment approach by trying out their own ideas.

When it came to the business of investing, my father offered me just a few pieces of direct advice. First, he said, make only the investment decisions about which you have a reasonably high level of conviction. It is a simple but often overlooked lesson: you cannot turn a profit if you are always second-guessing yourself. In and out traders usually enrich their brokers, not themselves. Second, cut your losses and cut them fast; do not listen to reason or emotion, just say good-bye. That was a lesson he had also learned from the 1920s.

This advice proved invaluable for the investing I was beginning to do then and for the building of the business that would come later. Also of great value was the discovery that you can look at the market, as well as the companies in it, in more than one way.

Understanding the cycles of the stock market was of great interest to my father. He liked technical analysis, an objective approach that involves the measuring and charting of market information over extended periods. (Technical analysis is probably the forerunner to today's more sophisticated total quality control; in both cases, the emphasis is on measurement.) Like my father, I was also intrigued by market cycles and psychological effects on the market. But my real interest was in what made one particular company good and another not as good, something called fundamental analysis. While money managers seemed to favor one form of analysis or the other, I realized very early that each discipline had enormous benefits as well as some shortcomings. I began to think that if you knew both, you would do better than if you knew just one. This belief soon affected my investing decisions and developed into the idea that the whole company should excel in both, with individual managers deciding which to use and when.

As an analyst, I visited many companies to evaluate their management and their earnings prospects. But I found myself more interested in the philosophy of how they ran the business and why they ran it in a particular way than the statistics I was being given. Over the years, I have learned a lot from studying other companies, which I have been able to apply to our business.

I also saw that sometimes the best stock buys are those that go against market psychology. Doing the opposite of what most investors are doing at any particular time is known as contrary opinion. I have found that being a contrarian works even better in business than in the stock market.

When my father turned over the day-to-day operation of the business to me in 1972, he again let me learn largely by doing. There were things I decided to do that I know he disagreed with, but he never interfered. All he wanted to know were two things: whether I was truly interested in the business and whether I shared his strong sense of personal responsibility to the shareholders— some of whom were relatives and close friends with substantial investments in the funds.

He had a certain confidence in me, despite my reputation in the family as the one who liked to spend money (a reputation that, in fact, has proven to be true). My father knew I could spend the money we made; he just did not know whether it would come back again. There was one year in the mid-1960s when the company made $5.5 million. I remember how guilty my father felt. He said, "A company like ours should never make this much money," whereas my attitude was, "This is wonderful! Boy, are we going to have fun with this money!"

To me business is fascinating. It is like a game; sometimes you win, sometimes you lose. There is no formula for success. Sometimes there are big rewards and sometimes no rewards; it is a matter of taking risks—the right risks, those where you believe the probabilities are in your favor. Shooting for the obvious profit seems like the best way to lose money because then you are always riding on someone else's coattails.

TAKING THE CONTRARIAN APPROACH

People have often asked me if it was a tough decision to get into the money market business. In fact, it was easy. It was a matter of survival.

Looking back at the fund industry, it had a wonderful period in the mid-to-late 1960s. Then, with the exception of money market funds, the industry basically died between 1969 and 1982. In the 1960s the fund managers at Fidelity could do no wrong. Everybody loved us. We were oriented toward supergrowth stocks—those with the best prospects of producing earnings quickly—and owned many secondary securities, which did well in the mid-to-late 1960s but poorly beginning in 1969. We went from being the most loved to being the most hated. Our sales fell about 95 percent. In the early 1970s, the assets we managed dropped from $5.5 billion to less than $1.7 billion. We were fighting for survival.

At the time, people were not interested in common stocks or stock funds, and two companies were ahead of us in launching money market funds. We needed to offer a service that was better and different. I thought that if we could provide investors with an easy way to take their money out through check writing, they would be more apt to put their money in. Some of my colleagues and competitors must have thought I was asking for trouble. But, in 1974, we introduced our first money market fund with check writing, called Fidelity Daily Income Trust, and we were surprised at how quickly the money started to pour in. We saw the money funds as a way to gather assets that hopefully would move into our stock funds once the market started to improve, giving us a head start on sales; we never saw them as a business in themselves. Dreyfus, which had launched a money market fund in 1972, promoted its product heavily and gathered the biggest share of assets. I learned from them that even if you have a good product, unless you promote it heavily, you will not gain market share.

A second thing we did that probably raised some eyebrows in the industry was to change our distribution. For thirty years we had sold our funds through broker-dealers. For a variety of reasons, that method of distribution was no longer working.

When I looked at the competition, I saw two strategies that gave me some ideas about how we could take a different approach to selling. First, I was impressed with how Dreyfus used clever marketing and advertising to create demand for its products. Second, Dreyfus had offered the first mutual fund that was aggressively managed for top performance. Fidelity's Capital and Trend funds followed the same style of investing. Our strategy was to buy stocks we perceived to be cheap and sell them when they became overpriced on a relatively short-term basis. The turnover was quite high, so it was considered somewhat scandalous at the time. But in the mid-1960s, the results were excellent. In both cases, we saw performance create sizable demand for the funds.

Many brokers believed that they were solely responsible for creating the demand for the funds, which may not have been totally correct. Investment performance and the investor's desire to make money was the mixture that produced explosive sales. It seemed to me that if our funds performed well, they would sell themselves. In some situations, we could eliminate the broker from the selling process and offer our funds directly to individual investors. That meant we could maintain a lower sales charge on the funds—or, in some cases, even remove it—because we did not have to pay commissions. Thus came our decision in the mid-1970s to sell our funds directly to individual investors through direct response advertising and a toll-free telephone line.

At the time, this was a radical departure from the way most funds were sold. Nobody else had tried to set up an operation that allowed customers to transact their business completely by phone. But we thought it would prove to be a cost-effective distribution system, and it has become just that. Nevertheless, a majority of people still prefer to buy funds with a full commission or sales charge. In fact, stock fund sales today are 60 percent full commission. Many people need the assistance of a salesperson to make their investment decisions. So we support and believe in both approaches—selling funds through brokers and selling them directly.

Despite the stock market's downturns in the 1970s, I remained optimistic because history was on our side. Looking back as far as the Civil War, I knew that the stock market was cyclical, and I had little doubt that it would start to move up again; I just did not know when. (I suppose if I had known in 1969 that we would have to wait thirteen years for another bull market, we might have moved into another business entirely!) At the time, we did not care about maximizing earnings or improving the income statement; we just did not want to lose cash. Remembering my father's experience in the 1920s and 1930s, I knew you could not run a business for long on borrowed money.

Business was so bad in the 1970s that it was actually a wonderful time to take over the company. There were no illusions about anything; we had to live by our wits. Based on that experience, I began to see stock market declines and economic recessions as opportunities to build the company and to make it stronger. They provided a chance for us to look at what we were doing and to try to do it better.

We decided in the 1970s to build and improve our investment department. In the mid-1970s, we did some major pruning and subsequently added to the group, even though our equity assets under management were not growing. Thus, when the bull market started up again in 1982, we had the talent to produce top-performing funds. I do not think that we would have been as well prepared if we had not laid the groundwork in the 1970s.

Another strategy was to bring the shareholder processing operations inside, a decision that added to the complexity of the business and may have caused people within the company to think that I was a little mad. Even my father had always thought that bankers knew best how to keep track of the books. If something went wrong, they would be responsible. It seemed to me that we were always at a great disadvantage if we were unable to communicate directly with our shareholders. If something went wrong and we said it was the bank's fault, the people who did business with us would say, "Well, who hired the bank?" Since we did, we had the ultimate responsibility when there was an error. A third-party outside servicing agent did not have the same stake in the success of the business as we did.

It had not gone unnoticed that Keystone—one of our Boston-based competitors—had a reputation for always providing good service to its broker-dealers and shareholders, largely because Keystone did its own servicing. It occurred to me that the major reason it sold a lot of shares was not investment performance so much as this high level of service. Of course, I knew from my experience as an investment analyst that successful companies usually had good marketing and sales departments as well as good products. To optimize both, it seemed to me we needed control over the service function in addition to the ability to produce top investment results.

Once we made the decision to develop shareholder servicing skills in-house, we did it one step at a time. The first function we brought in was that of the transfer agent, who handles shareholder record keeping; this includes tracking who owns shares of which fund and how many, when the shares were purchased, and what dividends are owed. Once we mastered that process, it seemed logical to next bring in-house the fund bookkeeping and accounting, which we did in 1973. By that time, we already had a history of gradually adding new skills and services. Before this, we had developed our ability to make money in stocks, then our ability to make money in bonds. After fully embracing shareholder servicing, we focused on learning to use the available technology effectively.

At times, our technology investments seemed contrarian. For example, although the company had little cash in the 1970s, we spent money on computers to enable us to handle our shareholder processing operations more efficiently. But it was not until the early 1980s that we started investing heavily in technology. Some of our competitors thought we spent excessively in order to have the latest technology. However, we viewed—and still view—the effective use of technology as essential.

We have tried to be timely in using new technologies, although I have found that if you are the first to use something new, you usually end up spending a lot of time working out the bugs. On the other hand, you cannot wait too long or your competition will get too far ahead of you, and you will never catch up. We have also tried not to fall in love with the technology itself. The technology may be fascinating, but we need to determine if it achieves the goal we want—quality service at a competitive price. In fact, the limiting factor today is not the technology itself but rather the training of people who are using it and writing creative software designs that do more than imitate old processing techniques. After all, technology is just a tool; the hand that guides the tool determines its performance. It is like putting a Stradivarius violin in the hands of an average six-year-old; he or she may love the instrument, but the chances of hearing beautiful music are very slight.

Technology has allowed us to do many things, such as build a two-thousand-person telephone sales and service force. But I believe we are still in the early innings of the game of learning what technology can do and how we can most effectively use it. The steam engine is an analogy. It was invented in 1825. Yet it was not until 1869—almost forty-five years later—that the transcontinental railroad was completed. Just as the steam engine changed the country by opening up new real estate, so technology will continue to revolutionize the way we do business. Any company that wants to be successful must, in my opinion, be part of that revolution; if it is not, another company will be.

Speaking of revolutionary, when we went into the discount brokerage business in 1979, everyone said I must be crazy. Mutual funds is a respectable business, they said. Why do you want to be in a business like discount brokerage, a business with a small profit margin? But there are advantages. Branch offices, for example, would allow us to build a distribution system with national visibility, something that would never be affordable with just mutual funds. And a brokerage system would allow investors to consolidate their stock, bond, and mutual fund investments into a single account. We could never offer that level of convenience with a mutual fund shareholder system. As it turns out, starting a discount brokerage business—we were the only mutual fund company to do so—was a good decision.

SWIMMING UPSTREAM

One of my distant nineteenth-century relatives once said, "Any fish can swim downstream, but only a live fish can swim upstream." This made sense to me as I thought about Fidelity's growth. We wanted to stay lively. We did not want to become entrenched in bureaucracy or to remain satisfied with our achievements. A company is constantly changing; it must be able to adapt. When a company stops changing and believes it has the secrets to everything, it has the secrets to nothing. Small companies know this. They are constantly changing. They face numerous challenges each day. To survive, they have to be nimble and creative.

Many companies, including banks, brokerage houses, and insurance firms, have diversified into mutual funds. But we were one of the first and only mutual fund companies to go into other businesses and to build them from scratch. Of the 1,500 mutual fund companies in the United States today, we are one of the only ones, for example, to start a discount brokerage service. Starting and developing new businesses has always been a key part of Fidelity's culture. Unfortunately, our first diversification efforts were venture capital investments in the 1960s that did not work out well. But this idea of Fidelity doing more than just managing mutual funds was something I had always wanted to do. And so we began diversifying into other businesses in earnest in the 1970s.

My father always felt that running an investment company was not a real business because it was "just a business of ideas." You pick up the telephone to call someone in; he buys the stock. You get in touch with the bank; it settles the trade. You call the bank again; it pays the dividend. Just ideas. It seemed to me that investing in other businesses could give us another possible service. In addition, each new business gave us an opportunity to try out some of our ideas and hopefully acquire new skills to help our basic business. Sometimes we tried to apply the skills learned in our basic business to start up derivative businesses, although we were not always successful.

In the early 1970s, for example, we decided to go into the mini-counseling business. I thought it was a wonderful idea because of the high fees associated with managing individual accounts. Most people did not manage individual accounts well, but I thought we could mechanize that with the help of computers. We bought a small company and put one of our portfolio managers in charge. The whole experience was a failure. We could not control the size of the accounts—they were all smaller than we had hoped. In addition, the investment results were bad. But we learned two important lessons.

First, we learned that it is nearly impossible to have a portfolio manager who can successfully handle individual accounts as well as deal with individual clients. A great investment manager is not necessarily an expert in customer relations.

Second—and this was before computers really were involved—we learned that bookkeeping impediments can kill a business. Keeping track of individual accounts, making sure all the right securities were in the right accounts, calculating the right percentage in each account—the mechanics of running the business—were a disaster. We lacked the necessary shareholder servicing skills.

So, the mini-counseling business failed. But it helped to push us to develop the skills we needed. In the 1980s, it seemed that the cost of venture capital deals had become too high. So we decided that instead of investing our excess cash in other companies, we would actively spawn our own small businesses. The name for these fledgling concerns is Fidelity Capital; among them are a car service, a chain of art galleries, a group of telecommunications companies, an insurance company, an executive search firm, and various publishing operations. The advantage that we have been able to offer the Fidelity Capital companies is an infrastructure that can supply capital, along with motivational and technical skills. These start-ups are free to use the knowledge and skills we have built into our basic business to help launch their efforts.

GROWING, LEARNING, CHANGING

People wonder why we build small businesses when we have such a strong basic business. There are many reasons, but the best one is building small businesses teaches you a lot about yourself and your organization. You keep alive, in an organization that has ever-increasing numbers of people, a skill that is essential—the ability to start something new and run it well. Starting out as an underdog and having to do a superior job is good practice for a growing company; the skills you learn can be applied to some of your "old" businesses to keep them alive and vibrant. Many of the businesses that we have developed can add something to Fidelity. What we learn in the telecommunications business, for example, can improve the marketing and sales department of our mutual fund operation, which depends heavily on computers and telephones. Plus, telecommunications is an industry with which we are familiar, so it makes sense to further our skills in that area rather than in an industry we know nothing about. Today it takes a whole group of skills—including investment, legal, technology, and general management—to make our business successful.

As Fidelity has grown, we have learned how important it is to work like a small company. One way we accomplish that is by continually breaking our basic business down into manageable units. We start a new company within Fidelity when we believe a particular business offers unusual opportunity for growth. Our 401(k) business, for example, which sells services for defined contribution plans to corporations, used to be part of our institutional business. In the mid-1980s, our institutional business had a fast-growing market selling mutual funds to bank trust departments. The 401(k) market, which was a much harder sale, was being ignored. So we broke it off and started a new company dedicated to that area alone. As the 401(k) business took off, it became apparent that the different markets it serviced needed to be treated differently—the needs of *Fortune* 500 companies with thousands of employees were quite different from those of small, emerging businesses. So we carved out separate subsidiaries to better serve each area. By breaking the business up into smaller units, people at all levels were able to take ownership and learn about the different aspects of a particular market in greater detail. We have found that if we do not carve our big businesses into separate small businesses, nobody takes responsibility for anything.

From our start-up companies at Fidelity Capital, we have learned that a new venture cannot simply be intertwined with the old. Since the old brings all the money in, and the new brings in none, a new business may not get the nourishment it needs. It is like trying to grow a little tree under a big tree—the little tree never gets enough sunshine. We can focus best on areas with the most opportunity by moving assets and people from other parts of the business. That way we can nourish the new area completely and hopefully make it to market with a better product faster than the competition.

To keep growing, we have to build new businesses and maintain and improve our old ones. In mutual funds, for example, we learned how to seed new funds and make them grow while keeping the older funds well-nourished so that they, too, produce competitive investment results. Whether starting a new, small business or running a part of the older, larger business, the basics are still the same. In either case, you have to understand exactly what will make the business successful.

With a small start-up business, we may try to do one or two things exceedingly well. With a bigger business, the goal might be twenty, thirty, or forty things done exceedingly well. In either instance, quality has to precede profit. But once we have succeeded in producing and maintaining a high-quality product or service, we have to make it profitable. Cutting corners, which may be a short-term solution to the profit part of the equation, hurts quality in the long run. No money is saved if quality is sacrificed; working smarter is what brings the two together. That means we have to be willing to try new ideas. My own rule of thumb is that a business has to be good for the customer (quality), good for the company (profitable), and good for the employees (rewarding). If we only achieve two out of three, we have not succeeded.

No matter the business, a good product is essential. That means determining what the customers want and then figuring out what we can provide. Over a long period, we have found that close communication with customers can mean the difference between success or failure. They give us the best feedback about our service; they tell us what they like or dislike about what we offer and what we do not offer that they wish we did. We try to respond to their requests whenever feasible, which is why building a spirit of change into the company is so important. It is also why paying attention to details is critical.

One way we have tried to do both is by implementing a Japanese philosophy known as *kaizen,* which had its beginnings in rice farming. *Kaizen,* a process that involves gradual but continuous improvement, has helped us build into the company the spirit that everything we do can be done better. Looking at the general quality of service that Fidelity provides its customers, it is fair to say we do a good job. But in closely examining some particular things that we do, especially in new areas, one finds that there are many rough edges. We are not as good as we think we are. *Kaizen* has helped us to strive for improvement when we might otherwise have been content with what we had already achieved. That is how we must continue if we want to remain a leader.

One of the best ways to become better at what you do is to learn how other people go about producing a better product or service. When Japan came out with superior products in the late 1960s and early 1970s, I wondered whether the difference could be attributed to a cultural factor or some process they were using. It fascinated me, for instance, that the Japanese could build cameras comparable to the German models at about one-third the cost. Over the years, I had read a lot about Japanese business practices. On a trip there in the mid-1980s, I picked up a book at the Okura Hotel called *Kaizen—The Key to Japan's Competitive Success* by Masaaki Imai. If *kaizen* worked so well in production facilities in Japan, I thought it might work just as well in servicing operations like ours.

We started with *kaizen*'s premise that how well we do little things daily could determine our success. If every employee strived to do his or her job better, that could translate into major improvements in how we as a company did business. The wonderful thing about making small changes is that you can see the effect they have on the total system. If a small change does not work well, it can easily be reversed. If it does work, then you can make another change, until cumulatively a tremendous change has been created. Of course, in order to detect improvement, objective measures of performance are needed for every aspect of a job.

In the spirit of *kaizen,* we continually look at what to measure, how to measure it, and what the results mean. Measuring the quality of what you produce in the service business is much more difficult than if you sell a widget that is supposed to last for ten years and breaks in ten days. So we measure a lot of things, from the quality of our investment results to the length of time it takes us to solve customer problems; once we have consistently met our benchmarks, we set new ones. Measurement is not an end in itself. It is a means to higher quality, timeliness, and lower costs.

Our company is good at coming up with products and services that customers like; making our businesses profitable is always harder and more time-consuming. My father started Fidelity in 1946, but the company never made a decent profit until 1960. Reaching profitability often takes perseverance. Still, businesses that do not turn a profit in good times are going to create losses in bad times. My opinion is that either we need to improve them so that they are profitable or leave them. To me, the willingness to trim businesses is as important as the willingness to build them. We do not want to trim them too soon, but if a business is not profitable, we need to look at it carefully to determine whether there is a real need for this service. If there is, we must decide whether someone will pay us enough to eventually make a profit.

The profit motive becomes important for the people who run our operations. Like my father, I believe in putting one person in charge and letting him or her go. If the person who runs a business is the right person—and, of course, if it is a good business—then there is a good chance of success. This is especially true of small start-up businesses. If you want a good entrepreneur—someone with singleness of purpose, love of the business, and absolute patience—you have to give him or her an entrepreneurial profit. The people who manage our start-up companies at Fidelity Capital have an agreement with us. If they build the business to a certain level, which we define in terms of sales and earnings, then they will receive a meaningful financial reward. The investment side of our business works the same way. We try to motivate people with a reward structure that is tied to producing results. Since Fidelity is a private company, the earnings can go to the people who produce the profit.

At all levels of the company, we have set up reward structures that encourage people to strive for improvements. Sometimes attempts at improvement fail; if they are sincere, intelligent attempts, the reward structure should accept that. On the other hand, we have to be careful that we do not go too far—experimenting with everything at once can spell trouble. We adjust one thing, prove that it works, and then move on to try other things.

We have been fortunate in finding good people and figuring out how to motivate them. Part of the motivation is financial, but part of it comes through making the jobs interesting and demanding. Pride is a factor. The most highly motivated, enthusiastic people take pride in their work, the same way a craftsman takes pride in the chair he has built. The realization that you are doing something important—something that adds value to yourself and to others—is a strong motivation. In fact, having talented and motivated people has helped us succeed more than any other single factor. Obviously, the best ideas in the world amount to nothing if you do not have the right people to execute them.

LOOKING AHEAD

Fidelity's biggest challenge these days is not fighting for survival; it is fighting against obesity. Stable companies run the risk of a metaphorical hardening of the arteries. But there are also dangers in running at high speed. When you grow fast, you work first on increasing capacity and then on improving the quality of the product; there is not time to develop real benchmarks. Yet those benchmarks are exactly what is needed to see improvements and to determine where you are making (or losing) money. Compared to a smaller enterprise, a larger company like ours has to be more organized, with good measurement and internal information systems. When we were small, our information systems were our personal relationships; as we grew bigger, we could no longer manage by gut feelings alone. Having the right measurements is critical now.

How big can Fidelity become? Who knows? A crystal ball might predict our involvement with a few more businesses. There are certainly many areas into which we could still expand—but only those businesses where we can do an excellent job. Whatever Fidelity looks like in the future, however, it will include quality products, quality service to our customers, and the satisfaction of the people who work here. We might not have the dominant market share in all our businesses, but it will be an important market share. What really matters is our ability to run our business exceedingly well.

Any company can fall into the trap of doing things badly—just look at how few companies have been around for twenty, thirty, or forty years. Every industry goes through growth phases and shakeouts. Many companies shrink or disappear after a shakeout; the survivors are the ones that can improve their products and cut their expenses. I live on a piece of land once owned by Sterling Elliot, who owned and ran Elliot Addressograph. The company was the leading automation firm in 1920—twice the size of IBM. It was also quite creative, and Elliot probably had as many inventions in the printing and automation area as any man in America.

But who today has heard of him or Elliot Addressograph? Both are totally forgotten. The lesson is that it is very easy to go backward—easier than going forward.

The element of personal satisfaction is critical for moving ahead. The business world is like a game. The question is not how well you play golf, for example; it is whether you enjoy playing it. You may be the best golfer in the world, but it means little if you do not play your best to beat your opponent. It is the same with business. Do the profits feel good or do they feel bad? And how about the losses? You can lose money without worry if the losses are part of the cycle and if the company is growing stronger. So it is a matter of balancing the desire to win against the sense of personal satisfaction you need to be enthusiastic about the job. If there was a simple formula for success and it was easy to follow, everybody would be doing it.

Who knows what the future holds? Bull markets or bear markets, I do not know, nor am I sure I want to. Certainly the power of computers opens up enormous new opportunities for processing and servicing in a wide range of businesses, including the financial industry. Asked what the future had in store in 1970, my father said, "Oh, heavens, you don't care about that. I guess my viewpoint on the future is very strong: I ignore it. It's so silly. You can't love the future. You can't hear it. You can't drink it. By thinking about the future, you suck all the vitality out of it. Aim to do whatever you're doing better and let all that take care of itself." I agree.

Index